Engineering Simulations as Scientific Instruments:
A Pattern Language

Susan Stepney • Fiona A.C. Polack

Engineering Simulations as Scientific Instruments: A Pattern Language

With Kieran Alden, Paul S. Andrews,
James L. Bown, Alastair Droop,
Richard B. Greaves, Mark Read,
Adam T. Sampson, Jon Timmis,
Alan F.T. Winfield

Susan Stepney
Dept. of Computer Science
University of York
York, UK

Fiona A.C. Polack
School of Computing and Mathematics
Keele University
Newcastle-under-Lyme, UK

Written in collaboration with Kieran Alden, Paul S. Andrews, James L. Bown, Alastair Droop, Richard B. Greaves, Mark Read, Adam T. Sampson, Jon Timmis and Alan F.T. Winfield.

ISBN 978-3-030-13202-6   ISBN 978-3-030-01938-9   (eBook)
https://doi.org/10.1007/978-3-030-01938-9

This Springer imprint is published by the registered company Springer Nature Switzerland AG
The registered company address is: Gewerbestrasse 11, 6330 Cham, Switzerland

Preface

Computer-based simulation is a key tool in many fields of scientific research. *In silico* experiments can be used to explore and understand complex processes, to guide and complement *in vitro* and *in vivo* experiments, to suggest new hypotheses to investigate, and to predict results where experiments are infeasible. Simulation is an attractive, accessible tool: producing new simulations of simple systems is relatively easy. But it is also a dangerous one: simulations (and their underlying models) are often complex, buggy, and difficult to relate to the real world system.

A recent UK Government report on computational modelling ([95], condensed as [44]) makes several recommendations, including

> Decision-makers need to be intelligent customers for models, and those that supply models should provide appropriate guidance to model users to support proper use and interpretation. This includes providing suitable model documentation detailing model purpose, assumptions, sensitivities, and limitations, and evidence of appropriate quality assurance.

In this book we describe the CoSMoS (Complex Systems Modelling and Simulation) approach, a pattern-based approach to engineering trustworthy simulations: simulations that are both scientifically useful to the researcher, and scientifically credible to third parties. The CoSMoS approach emphasises three key aspects to this development of a *simulation as a scientific instrument*: the use of *models* to capture the scientific domain and the simulation platform; the use of *arguments* to provide evidence that the scientific instrument is fit for purpose; and the *close co-working* of domain scientists and simulation software engineers.

The CoSMoS approach is generic: it does not mandate a particular modelling technique, a particular implementation language, or a particular real world application domain. What it does mandate is the careful and structured use of models and arguments, to ensure that the simulation is both well-engineered, and seen to be well-engineered.

The book is useful for domain scientists who wish to see what is needed to build and use scientifically credible simulations, and for software engineers who wish to build scientific simulations that are useful and usable. The examples in the book focus mainly on biological simulations, to exploit the specific experience of the authors, but the CoSMoS approach is not restricted to this domain. The approach we describe is also useful for building simulations as part of the process of building engineered systems. There are some differences between scientific simulation (crudely, simulating the world as it is) and engineering design simulation (simulating the world as we wish it to be), but much of the approach is common. These differences are noted when they occur.

What this book is not

This is not a book about software engineering. Where we advocate particular software engineering approaches and good practices for building simulations, we mention them in the endnotes[1] and refer to standard literature discussing them.

Neither is this a book about specific modelling approaches or notations[2], whether mathematical, computational, or otherwise. Again, there is much excellent literature available on specific modelling techniques, to which we refer.

Nor is it a book about experimental methods, although the simulations are used to perform (numerical, virtual) experiments in a way analogous to real world experiments.

Nor is it a book about developing bio-inspired algorithms (which are abstractions of biology, but not simulations of biology).

What this book is

This is a book about how to combine well-established approaches in a manner that leads to a robust and argued-trustworthy simulation: how to combine modelling to build an appropriate scientific instrument, software engineering to build the instrument appropriately, and experimental techniques to use the instrument appropriately for scientific investigation.

Part I provides a managerial overview: the rationale for and benefits of using the CoSMoS approach, and a small worked example to demonstrate the approach in action.

Part II is a catalogue of the core CoSMoS patterns. Start your project at the CoSMoS Simulation Project (92) pattern, and follow the guidance, using the other referenced patterns as your specific context demands. The development and use of a simulation as a scientific instrument will probably include most of these core patterns.

Part III is a catalogue of "helper" patterns. These are more specific patterns of use only in certain domains, for certain purposes, or with certain modelling and implementation approaches. Use of these is not required to develop a CoSMoS simulation, but they provide possible routes for doing so.

Part IV documents CellBranch, a substantial case study developed using the CoSMoS approach.

Acknowledgements

The original CoSMoS (Complex Systems Modelling and Simulation) research project (2007–2012) was funded by the UK's Engineering and Physical Sciences Research Council (EPSRC), grant numbers EP/E053505/1 and EP/E049419/1, with additional support from Microsoft Research and Celoxica. The CellBranch research project (2014–2015) was funded by the UK's Biotechnology and Biological Sciences Research Council (BBSRC), grant number BB/L018705/1.

A list of CoSMoS publications and associated tools is available in the endnotes[3].

The York Computational Immunology Lab (YCIL) is using CoSMoS for many of its projects. It has made extensive material available, including simulations, simulation analysis tools, and fitness-for-purpose argumentation tools, available at www.york.ac.uk/computational-immunology/software/.

The author team have made substantial and various contributions to the CoSMoS project, to the CellBranch project, to YCIL projects, and during all stages of the development of this book. In addition to their overall work on the structure, writing, and editing of this book, they have had the following specific roles on the projects, and, as far as can be determined from such a joint effort, have made the following specific contributions to the book:

- Susan Stepney (Principal Investigator of CoSMoS and CellBranch): developer of the detailed pattern language structure

- Fiona Polack (Co-Investigator of CoSMoS): developer of the detailed argumentation pattern suite
- Kieran Alden (Research Associate applying CoSMoS in YCIL): contributor to Calibration and Sensitivity Analysis patterns
- Paul Andrews (Research Associate on CoSMoS): developer of the initial CoSMoS philosophy, process, phases, and high-level models: contributor to Chapter 1
- James Bown (advisory member of CoSMoS): contributor to Chapter 2
- Alastair Droop (owner of the AlKan worked example in Chapter 3): contributor to Chapter 3
- Richard Greaves (Research Associate on CellBranch): developer of Part IV
- Mark Read (then PhD student applying CoSMoS in YCIL): contributor to Calibration and Sensitivity Analysis patterns
- Adam Sampson (Research Associate on CoSMoS): contributor to documentation and real world patterns
- Jon Timmis (Co-Investigator of CoSMoS; applying CoSMoS in YCIL): contributor to Chapter 1
- Alan Winfield (advisory member of CoSMoS): contributor to Chapter 2, embodied simulation pattern, and several antipatterns

We thank all our other CoSMoS project partners, CoSMoS workshop contributors, CoSMoS case study domain scientists and simulation engineers, and in particular Christopher Alexander, Frederick Barnes, Sabine Dietmann, Fiona Frame, Philip Garnett, Teodor Ghetiu, Julianne D. Halley, Tim Hoverd, Norman J. Maitland, Carl G. Ritson, Austin Smith, and Peter H. Welch.

Susan Stepney, Fiona Polack
York, UK, 2018

Contents

Part V Appendices

Part I
Overview of the CoSMoS approach

The first part of this book provides a managerial overview: the rationale for and benefits of using the CoSMoS approach, and a small worked example to demonstrate the approach in action.

Chapter 1
CoSMoS: rationale and concepts

Abstract — In which we discuss: why CoSMoS is needed; simulation as a scientific instrument; the use of a pattern language to define CoSMoS; interdisciplinary collaboration; and the underlying concepts of the CoSMoS approach to complex systems modelling and simulation.

1.1 Why CoSMoS

Biological systems, social systems, the entire planetary system: these are complex systems comprising large numbers of components of diverse kinds, with complex behaviours, non-linear dynamics, and non-intuitive emergent properties. Computer simulation provides a mechanism for exploring such systems. In this context, a computer simulation is an executable form of a model of a system, used to animate the dynamics implicit in the model, and determine its consequences.

Systems that have mathematical models based on well-understood physical theories can be 'solved', occasionally analytically, more usually numerically. However, complex systems typically do not have such a basis of well-understood theories. Models of such systems typically have a "nest-like" structure [151], weaving together data and sub-models from multiple sources, incorporating hypothesised components and mechanisms, all forming a more-or-less robust structure. With such a nest-like, complicated, and potentially opaque model, it is essential to have an open and rigorous process for developing and using simulations, in order for there to be any confidence in the results, and to allow the results to be challenged in a principled manner.

© Springer Nature Switzerland AG 2018
S. Stepney, F.A.C. Polack, *Engineering Simulations as Scientific Instruments: A Pattern Language*, https://doi.org/10.1007/978-3-030-01938-9_1

Before embarking on any specific use of simulation, it is worth remembering that a computer simulation is only as good as its underlying model and data "nest". A simulation is exercising the *model*, not *reality* [60]. Any prediction it makes about the consequences of its model should be tested against the real world [140]. We advocate using simulation to *test the model*: if model predictions and real world observations diverge, the model needs to be revised. This may require new real world experiments, to provide new data and structures to update hypothesised components of the nest-like model. Such use of simulation forms a crucial part of hypothesis-driven research life-cycle in, for example, systems biology [141]. In this case, it is essential to have a rigorous process for updating and clarifying the underlying models, and ensuring the simulation is faithful to the specific model under test.

CoSMoS, Complex Systems Modelling and Simulation, provides such a process.

The rest of this chapter describes the conceptual basis for CoSMoS: the need for a set of rigorous *models* developed for a given scientific *purpose*, and an associated fitness-for-purpose *argument*, defined using a *pattern language*. Chapter 2 gives further motivation for CoSMoS from two distinct application domains: systems biology, and robot development. Chapter 3 gives an overview of the entire CoSMoS approach in miniature, by use of a small example development, tailored to illustrate the core CoSMoS components. Parts II and III form the core of the book, detailing the CoSMoS pattern language; Part IV is a substantial case study illustrating the use of the entire CoSMoS approach.

1.2 Complex systems simulation

We use the term *model* to refer to any abstraction that is made to explicitly describe or define concepts for the purpose of communication, documentation and/or understanding. Such models are typically static and aim to capture structures and behaviours within and among their subject components, or data relating to those components.

This book is concerned with constructing and using "mechanistic" style simulations, encoding a logical model of hypothesised behaviour in order to execute and explore that model. Such a simulation can be interrogated and observed, providing insight into the model, and the subject of the model.

Simulation can support scientific exploration in numerous ways including communication, theory exploration, hypothesis generation, and design of real world experimentation. This is best achieved through close collaboration between scientists and simulation engineers, who together shape the purpose and use of simulation. The ideas laid out in this book provide an approach to augmenting the traditional experimental scientific approach through engineering a bespoke instrument, the simulation, in order to contribute to the understanding of the system of study. Consequently, we must have confidence that the simulation can actually tell us something about the real domain.

CoSMoS addresses the issue of trusting simulation outputs by providing advice that makes explicit how the simulation platform has been engineered and used to generate its outputs. This exposes the work to review and challenge, helps to provide scientifically reproducible results, and maximises the impact of simulation in science. It is the duty of the simulation engineer to make the tools they develop as accessible as possible without trying to deceive: just because the simulation results look like the system being modelled, does not make them correct, and certainly does not mean that they can be extrapolated to hold in unstudied real world scenarios. Care is required to enable the scientific audience to understand: how to use the developed simulation technology; how it works and upon which abstractions it is built; and how to apply its outputs to real world systems of study. Higher levels of trust are more expensive: they require more rigorous development practices, and more rigorous argumentation. The level of trust demonstrated for the simulation should be appropriate to the criticality of the simulation, which largely depends on how the results are to be used.

1.3 Simulation as a scientific instrument

We take the view that computer simulations are in many ways a kind of *scientific instrument*, like those used throughout science. Hence computer simulations should be subject to the same rigour that goes into constructing other kinds of scientific instrument. They need to be calibrated to understand how the outputs relate to the system under study, and they should be presented in such a way that their findings can be reproduced.

To this end, we distinguish the Simulation Platform, being the simulation code that can be used as an instrument to explore the relevant model, from the Simulation Experiment, being the use of that instrument to perform a specific investigation.

Simulations are based on an underlying model used to represent the system or domain under investigation [79, 123]. Simulation results are driven by this underlying model, and are not (directly) connected to the physical reality being modelled [60]. However, that is increasingly the case with many of the more sophisticated scientific instruments in use today, from fMRI medical scanners to the LIGO gravitational wave detectors: measured results undergo significant processing either in the instrument, or subsequently, based on some underlying (and potentially opaque) model. So simulation as a scientific instrument can be thought of as being at one end of a spectrum of distance from reality, not as a complete difference in kind from other scientific instruments.

Simulations are built for a *purpose*; there are many different purposes. Galton [83] outlines four potential uses for simulation: *prediction* (run the simulation from present data to predict the future), *explanation* (run the simulation from past data to explain the present state), *retrodiction* (run the simulation from hypothesised past data to test the hypothesis), *planning* (run the simulation from a planned near future state to test if it results in the desired goal). Epstein [65] lists 16 reasons to build a model beyond that of prediction, including discovering new questions, demonstrating whether the model conforms to reality, and for training and education. For example, one may construct a simulation to investigate the emergence of observed natural phenomena from hypothesised underlying behavioural processes. Such simulations may explore scientific questions about real world physical, biological or social systems; the results may be used to inform real world experimentation on the system being investigated, to validate such experimentation, or simply to explore concrete or abstract hypotheses. Different purposes will result in different simulation instruments being built: the purpose must be clearly articulated.

Learning takes place during both construction and manipulation of the model that underlies the simulation. In the simplest scenario, model construction results in the simulation computer code, and model manipulation takes the form of *in silico* experimentation. During the former we learn about the system and gain an idea of the questions we wish to ask of it; during the latter we explore these questions and enhance our understanding of the model upon which the simulator is based.

The construction of scientific instruments, like simulations, is based on a model of our understanding of a target system. For example, in order for an optical telescope to be constructed, we need to understand how light refracts using lenses. Here the model is embedded within the physical construction of the lens, and by mechanical adjustment of the instrument, the human observer is presented with magnified images of the telescope's target. Compared to traditional scientific instruments, the model of understanding embedded within a computer simulation is almost entirely logical and achieved in software. Additionally, input from the domain of investigation to a simulation is far more indirect with regard to space and time in the sense that no direct physical input is present. Again, we rely on logical connections to the domain rather than direct physical inputs. This results in an additional layer of interpretation required to understand how the inputs of computer simulations map to the entities of the domain under investigation.

Understanding the relationship between an instrument and the domain it measures is essential to interpreting its output. To achieve this, an instrument is *calibrated* to produce outputs that are meaningful to the human observer. Calibration relies on correctly observing and reproducing the structure of known features measured by the instrument [123]. Simulators are essentially unique *bespoke* instruments constructed to answer a specific question or set of questions. The purposes of these instruments are various, and therefore calibration is different for each individual simulator. When changes are made to a simulator it may have to be re-calibrated.

Rigorous calibration is a first step towards achieving scientific reproducibility, the need for which is becoming apparent within scientific simulation. Timmer [216] describes the beginning of a movement towards researchers adopting approaches to ensure that computational tools are in line with existing scientific methods. However, whilst there may be a recognition that nearly everyone doing science uses some form of computation, there are few who know what is needed to make sure that documentation of approaches is sufficient for reproducibility.

Exact reproducibility of a piece of science performed using computer simulation without access to the computer code and all the initialising variables (parameter settings, initial states of data) is unlikely. This gives support to the argument for complete openness of code [125]. There is another equally crucial source of knowledge that should also be open. The simulation encodes a model. This model may be implicit: expressed only via its encoding as computer code. It contains many different assumptions which are essential to understanding what the model represents. This is an issue of *validation* [201][1]: how you know that you have built the right system for your purpose. This can only ever be expressed as a level of confidence. In some circumstances, for example where outputs of a simulation instrument have a high level of criticality, a structured argument is required to express confidence in a computer simulation [181], and for this, the model must be explicit.

In the aftermath of 'Climategate' [111], the spotlight is more closely on the way in which scientists use computational devices as part of their scientific process. There have been calls for open-source code to enable repeatability, but in our view this is not enough. We need also to make the models and their assumption open, the calibration open, and openly argue the fitness of a simulation for a particular purpose. We need to demonstrate *how* the simulation has been engineered, and *why* it is a good instrument to enhance our domain knowledge.

In biological and other scientific research publications, "methods and materials" are a prominent section in any experimental paper, describing how relatively well-known and commonly used processes are pieced together for the described experiment. Rarely does any analogous material find a place in simulation literature (the ODD protocol [103, 104] is a welcome exception), and certainly not at the same level of detail. And because in simulation much is bespoke, there is less scope to rely on pre-existing and well-accepted "modules" in describing the work.

CoSMoS provides a framework for making all these aspects open, and documentable, at the appropriate level of rigour.

1.4 Simulations in engineering

Simulation also plays an essential role of simulation in engineering and technology. There are many applications of simulation technology across engineering research, design and operation, and more broadly for training and entertainment (video games). Here we outline a number of illustrative examples. CoSMoS can also be used as an approach for building such engineering simulations.

1.4.1 Training and video games

Perhaps one of the earliest applications of simulation technology was as part of professional flight simulators. From the 1960s digital computers have provided the computation, typically combined with a realistic cockpit mounted on a motion (Stewart) platform, and visual displays providing the pilot(s) with the view from the cockpit windows. A flight simulator is required to model, to a high degree of accuracy and in real-time, the flight dynamics of the aircraft and its systems (including engines). In response to the flight controls, the simulator must generate realistic values for the cockpit instruments and update the Out-The-Window (OTW) visual scene. A characteristic of the flight simulator is that it is designed to be immersive; a flight simulator is thus an example of immersive simulation environment more often known by the term Virtual Reality [190].

Even for video games that do not provide an immersive environment, the computational and graphical performance of personal computers and games consoles provides a level of performance, in respect of modelling the

OTW view, that approaches that of professional training simulators. Flying and driving simulators represent a class of video games noted for realism, to the extent that a winner of a racing simulation video game contest secured the chance to race for real, achieving creditable performance on the basis of skills apparently learned from the video game Gran Turismo [97].

1.4.2 Engineering Research

Simulation has become an essential tool in robotics research and development. Typically a robot simulator has four key features: (i) it provides a virtual world, such as a 3D model of the real world, modelling physics, collisions, etc; (ii) it provides a facility for creating a virtual model of the real robot, modelling its structural components, actuators and sensors with reasonable fidelity; (iii) it allows the robot's software controller to be installed and 'run' on the virtual robot in the virtual world, and ideally it is possible to install and run the same control code on the real robot; (iv) it provides a visualisation of the virtual world and the simulated robots (and other objects) in it.

In a well-designed robot simulator we can develop and prototype new robot designs (both hardware and software), research new robot control algorithms, including algorithms for multi-robot systems, and, by placing the simulator within a Genetic Algorithm framework, artificially evolve new robot designs.

Well known robot simulators include the open source Player/Stage tools [87] and the commercial simulator Webots [162]. These are examples of so-called sensor- and physics-based simulators, in which a reasonable trade-off between the accuracy with which robot sensors and actuators, and their interactions, are modelled and run-time speed of the simulation has to be achieved. The "reality gap" between the simulated model and the real world represents a challenge to robotics researchers, and is discussed in detail in §2.3.1.

1.4.3 Engineering and Process Design

In electronics design the use of simulation tools for both analogue and digital electronics circuit design is very well established. The well-known electronics circuit simulation SPICE (Simulation Program with Integrated Circuit Emphasis) [226], and its many variants and derivatives, performs AC and DC analysis, noise analysis, transient analysis and transfer function (input-output) analysis. Integrated circuit manufacturers make use of such electronic circuit simulators to validate designs prior to manufacture – in some cases with sufficient confidence that costly fabrication runs for prototyping and test are unnecessary.

An important class of immersive VR is the CAVE (Cave Automatic Virtual Environment); typically a room with video projected images onto walls, floor and ceiling, and a motion capture system to track the user's movements so that the CAVE display can respond to those movements. CAVE systems are used in a wide range of applications, including product development (to, for instance, virtually test a user interface) or process engineering (to virtually prototype a factory layout).

1.4.4 Engineering Operations

Where engineering plant and machinery is remote, or the consequences of an incorrect control command are potentially catastrophic, then simulation can provide a tool for testing and validating control actions before those actions are committed by the real system. An example of such remote systems are the Mars Exploration Rovers, Spirit and Opportunity; before commands are sent to the vehicles on Mars, they are tested and validated using computer simulation tools and scenario simulation on duplicate vehicles in a terrestrial "Mars yard" environment [129].

An example of a simulation that is very closely coupled with the real plant and machinery is the robot arm within the Joint European Torus (JET) facility at Culham, Oxfordshire. A multi-axis manipulator within the torus allows engineers to conduct maintenance or repair, by remotely controlling the robot arm and its tools. Several cameras within the torus, including cameras fitted to the robot arm, provide views to the operators but these, not surprisingly, provide only limited viewpoints. Relying on these cameras alone creates a significant risk that, in manoeuvring the robot's end-effector, some other part of the robot arm (an elbow for instance) collides with the critical inner skin of the torus. The solution is a 3D simulation of the robot arm, together with the structures of the torus, that is coupled to and run in parallel with the real system. In the control room the simulator visualisation is displayed alongside the camera views from the real system. Each time a new movement of the robot arm is planned, the simulated robot arm is reset to exactly match the disposition of the real robot arm; the next planned action is then run on the simulator. Importantly the simulator provides views to the operators from any desired viewpoint –even from outside a (transparent) simulated torus. Only when the operators are satisfied that the action poses no risk is it carried out for real [200].

1.5 Working together

Understanding and control of complex systems is a significant motivation for collaborative research. Complex systems require research across conventional domains, and the lack of knowledge about how a particular domain is involved in and with the wider context is a common limiting factor.

The point of collaboration is not to become an expert in someone else's domain, but to understand how to add one's own expertise to an attempt to address a problem expressed in a shared context. Expertise could be seen as the right to pontificate about a particular domain, but it also entails an obligation to conduct domain research to a high standard.

Collaboration is usually serendipitous: people or groups from different domains build on chance encounters to take forward research that the individual or group could not pursue alone. However, with the increasing emphasis on collaborative and interdisciplinary research, the possibility of constructing collaborations is increasing. What would an ideal collaboration look like? The following are some pointers:

- skilled scientists and engineers from more than one domain
- mutual commitment to a shared goal
- mutual trust in the expertise (as identified above) of other parties
- individual participant commitment to excellence in their domain of expertise
- a problem that cannot be addressed within one domain

We are concerned with creating simulations that are fit for purpose in scientific and engineering research domains. In this context, there are necessarily research domain experts, and software engineering domain experts.

Arguably, an attempt at complex systems simulation without expertise in both domains cannot be shown to be fit-for-purpose. The engineering of software is not simply a matter of building a program with a visual interface, any more than understanding the research domain to be simulated is simply a matter of access to an undergraduate textbook or a few scientific papers.

In systems analysis, the concept of a *role* provides a useful generalisation: a role conveys rights and obligations; a role is played by one or more people, and one or more roles can be played by a person. A role-based consideration of collaboration helps to identify the actual areas of expertise required. Figure 1.1 summarises roles in the context of research simulation.

Role	Description	Contributes
Domain Scientist	one or more scientists who commission the simulation with the intention of using outputs from the simulation in their domain research	scientific purpose and scope; indication of impact and criticality; information about the domain; advice on and interpretation of scientific sources; factual input and data; advice on calibration experiments and sensitivity analysis
Domain Modeller	one or more software engineers who collaborate with the domain scientist to create a conceptual model of the domain to be simulated	a model using concepts and notations that the domain scientist can recognise and understand, but which also admits a systematic route to implementation
Simulation Engineer	one or more software engineers who collaborate with the domain modeller to engineer a simulation derived from the domain model; and collaborate with the domain scientist to test and calibrate the implemented simulator, and run simulation experiments	software design and implementation in which appropriate domain concepts are traceable, and which includes facilities to set up and run simulation experiments, with appropriate observation and data collection
Argument Modeller	one or more modellers responsible for collating the basis on which the other roles can reach consensus on the fitness for purpose of the simulator	a fitness-for-purpose case that exposes the rationale for trust by all roles in the simulator, the assumptions made, the limitations in the purpose and scope of the simulator and in the potential use of its results

Fig. 1.1 Example roles in development and use of a collaborative research simulation

Not everyone can work collaboratively. The people playing each role need flexibility, and commitment to the simulation project. Everyone needs to accept the challenge of explaining their area of expertise in non-specialist terms: all parties need to understand that they can challenge the other experts involved, ask questions, seek clarification and explanation. Because it is a complex systems project, it is also necessary that all participants understand that "we do not know" or "we cannot possibly know" are valid answers to many questions about the domain (e.g. what size is particle x; how many of y are produced), the domain modelling (e.g. is a spatial model compatible with the scientific data; how can we tell if we got it right) and the implementation (e.g. what is the effect of digitisation; how random is random; is behaviour x realistic, a feature, or a bug).

The human aim of collaboration is to arrive at a consensus on the fitness for the scientific purpose of the simulator. The participants must not take for granted other people's opinions, but should work constructively together to understand assumptions and limitations.

The people involved in each role need to see the simulator as a tool: a scientific instrument that is only as good as what goes in to its construction. The use of the simulator and simulation results depends on this understanding.

1.6 Pattern language

1.6.1 Historical context

We document the CoSMoS approach using a *Pattern Language*.

In 1977, Christopher Alexander and his co-authors published *A Pattern Language* [7], one in a series of books "intended to provide a complete working alternative to our present ideas about architecture, building, and planning". It is a handbook of 253 patterns, where "Each pattern describes a problem which occurs over and over again in our environment, and then describes the core of the solution to that problem, in such a way that you can use this solution a million times over, without ever doing it the same way twice." [7, p.x]. The patterns describe how quality buildings should be designed, and together provide a language covering a wide range of spatial scales, from whole towns, through small clusters of buildings, and individual buildings, to tiny detailing.

And that, as far as the computing community goes, would have been that, were it not that the concept of Patterns inspired a group of software engineers. Buildings are not the only things described by "architecture": software engineering uses the same word to describe its own structuring concepts. In 1995, the so-called "Gang of Four" published *Design Patterns* [84], which took Alexander's concept and used it to produce a catalogue of patterns found in good software architectures. Things have not looked back: there are now analysis patterns [74], coding patterns [26], HCI patterns [225], patterns conferences and catalogues [56, 112, 156, 223], antipatterns [39, 143], metapatterns (patterns that describe patterns), and more (including arguments that the whole software patterns community has completely missed Alexander's point [81]).

The flurry of publications may have slowed somewhat since those early days [224], but Patterns are now part of the everyday culture of software engineering. One impact of Alexander's ideas, as adapted by the Gang of Four, on software development has been to make it clear that there is much more to object-oriented architecture than just the single concept of an object. The patterns provide a simple vocabulary, letting us all talk of the Visitor Pattern, or the Factory Pattern [84], without having to explain what we mean.

A pattern *language* provides a collection of associated patterns that can be used together to achieve a purpose. The CoSMoS pattern language provides a collection of patterns for designing, building, using, and arguing fit-for-purpose, simulations as scientific instruments.

1.6.2 CoSMoS pattern structure

It is important that a pattern is a practical, tried-and-tested solution to a problem, not merely something the pattern writer hopes or theorises might be a good solution[2]. The CoSMoS pattern language has been used on several simulation developments, and we provide references where appropriate.

We use the following template to document a pattern:

Pattern Template

Intent

What the pattern is for; what its use will achieve.

Summary

The tasks (some stated as patterns) needed to achieve the pattern's intent.

Context

The place or circumstance where the pattern is applicable.

Discussion

A general discussion of what the pattern provides, why it is needed, and how to use it, including examples of its use where appropriate.

Related patterns

A discussion of related patterns, and of antipatterns to beware of when applying this pattern.

We refer to a pattern in the text by its name in sans serif face and the page number where it is documented – Pattern Template (14).

A full pattern language is more than just a vocabulary. Patterns can join together to form larger patterns; the use of one pattern leads to the use of another; antipattern solutions suggest alternative approaches. Many CoSMoS patterns are expressed as a composition of smaller, more detailed patterns. This suggests a structure to the final CoSMoS project artefacts, but should not necessarily be taken as a suggestion on the *order* of pattern use. The development process should be flexible and iterative, and tailored to the specifics of a given project.

1.6.3 CoSMoS anti-pattern structure

Patterns provide guidance on what to do. It is just as important to give guidance on what *not* to do, particularly when this superficially appears to be a good idea, a clever shortcut, a sensible compromise, or even just normal practice. Antipatterns [39, 143] provide a means to give such guidance. An antipattern documents a pattern of bad behaviour or an often repeated mistake, together with a solution of what to do instead, or how to recover from the mistake. The solution is often a pointer to which pattern(s) to use instead.

We use the following template to document an antipattern:

Antipattern Template

Problem

What the problem is.

Context

The place or circumstance where the mistake is often made.

Discussion

Further discussion of the problem.

> ### Solution
>
> *A pithy summary of what to do instead, or how to recover from the mistake.*

We refer to an antipattern in the text by its (usually pejorative) name in italic sans serif face and the page number where it is documented – *Antipattern Template* (15).

In addition to antipatterns of the form "doing the wrong thing", antip-atterns can often appear in pairs (for example, *Analysis Paralysis* (138) and *Premature Implementation* (170)) where one of the antipatterns is "doing too much" and its pair is "doing too little".

1.7 The CoSMoS concepts

CoSMoS assumes the use of good software engineering practices, but this book is not about specific software engineering techniques. Instead, CoS-MoS provides a pattern language (§1.6) that enables the construction and exploration of simulations for the purpose of scientific research: it defines where good modelling and software engineering techniques should be used. This patterns-based approach has been designed to be adaptable both to a variety of simulation problems and to changing circumstances during sim-ulation construction and use. Application of appropriate patterns should be tailored to suit the criticality and intended impact of the research outcomes.

The construction and use of simulations is a necessarily interdisciplin-ary endeavour between scientists who study a particular Domain (123) (tak-ing the role of Domain Scientist (101)), and software engineers who construct simulations to facilitate the study of that domain (taking the roles of Domain Modeller (103) and Simulation Engineer (105)). The Domain Scientists, Domain Modellers, and Simulation Engineers are involved together in open-ended sci-entific research: the simulations are used as a tool to support theory explor-ation, hypothesis generation, and design of real world experimentation.

To run computer simulations we need to engineer a Simulation Platform (161). A properly calibrated Simulation Platform is the *scientific instrument*, the basis for running a Simulation Experiment (177). To engineer such a platform requires us to explicitly represent some knowledge of the system being stud-ied in a form of source code that can be run on a computer. This source code is either designed manually by the developers or automatically generated from a higher-level description.

In many existing approaches to simulation, the source code is the only explicit description of the aspects of the target domain that are being simulated. Source code contains numerous implicit assumptions[3] concerning both the scientific aspects of the work, and the engineering design of the simulation platform. Source code also contains many implementation details, which are needed to make the simulation run on a computer, but are not part of the underlying scientific model. Hence source code is not a satisfactory model of the target domain.

To mitigate inappropriate assumptions in the design of a Simulation Platform and to have greater confidence that simulation results can actually tell us something that relates to the real system being studied, CoSMoS proposes a series of related models to drive and describe the development of the Simulation Platform and simulation results generated from its use. Systematic development assists interaction between Domain Scientists and software engineers, and improves confidence in, and interpretation of, the results of simulations.

1.7.1 Phases

We identify three main phases in a simulation project.

Discovery Phase (95) *Deciding what scientific instrument to build:* establishing the scientific basis of the project, including the scientific purpose addressed by the simulations; identifying the domain of interest, models the domain; shedding light on scientific questions.

Development Phase (96) *Building the instrument:* producing a simulation platform to perform repeated simulation, based on the output of Discovery.

Exploration Phase (97) *Using the instrument to run experiments:* using the simulation platform resulting from Development for exploring the scientific questions established during Discovery.

These phases are not intended to be performed purely sequentially. A project naturally begins with discovery, followed by development and then exploration. But many iterations of discovery, development and exploration may be required to build a robust, fit for purpose instrument. The separation into phases helps provide a focus on what particular pieces of information are needed at each phase for each model.

Indeed, some projects might not perform all phases. A prior project may have performed the necessary discovery, and only development and exploration is needed (although it will be necessary to check that the assumptions of the prior discovery phase are valid for this project). Similarly, a suitable existing simulation platform might exist, and only the exploration phase is

Fig. 1.2 Relationship
between CoSMoS Simu-
lation Project components;
arrows represent flows of
information. These are all
framed by the Research
Context (119)

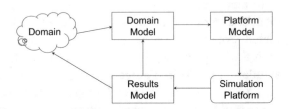

followed in this project (again, it will be necessary to check that the assumptions underlying existing the simulation platform are valid for this project).
On the other hand, it may be that only the discovery phase occurs, and dis-
covers that a simulation is not appropriate, or not needed.

1.7.2 Models

Our simulation approach uses the following model concepts: domain, domain model, platform model, simulation platform, and results model (Figure 1.2). Each of these components has a different role to play in the building,
verifying, and use of the simulation:

Domain (123) A particular view (or views) of the real world system of study.

Domain Model (128) A descriptive[4] (scientific) model of the Domain, as understood from domain experiments, observations, and hypotheses of
underlying mechanisms. It encapsulates the understanding of appropriate aspects of the domain. It focuses on the scientific understanding; no
simulation implementation details are considered.

Platform Model (149) A prescriptive (engineering) model of the Simulation
Platform. It comprises computational analogues of relevant Domain Model
concepts, and includes implementation details needed for the Simulation
Platform.

Simulation Platform (161) An encoding of the Platform Model into a software and hardware platform. It is the platform with which Simulation
Experiments can be set up, performed, and observed.

Results Model (174) A descriptive model of the *simulation* domain, as understood from Simulation Experiments and observations. It encapsulates
the understanding that results from Simulation Experiments, and casts it
in Domain Model (128) terms to support comparison with Domain Model
(128) analogues.

Underlying these models is a common Data Dictionary (132), defining the
modelling data used to build the simulation, and the experimental data that

is produced by domain experiments and corresponding simulation experiments.

CoSMoS does not dictate any particular form of modelling. The important things to consider when deciding what sort of models to create are:

- does the Domain Scientist understand the models of the Domain? can the Domain Scientist tell when the Domain Modeller has got something wrong?
- can the key concepts of the Domain and the Simulation Platform be expressed in the models? do we need to create a domain-specific interpretation of a modelling language, or use different languages for different views?
- is there, or can we devise, a clear mapping from the domain model to the platform model and to the code of the simulation platform?

That is, the key point is to select modelling approaches that all parties are comfortable with, and that are suitable for expressing the domain model. Put effort into traceability, all the way from Domain Model through code to results, rather than adjusting a model to suit the software engineering.

In §1.7.4 we describe the common substructure of the three main models. In Chapter 3 we illustrate the entire approach in miniature, through a small running example.

1.7.3 Implementation and calibration

The models described above are used to specify, design, and build the Simulation Platform. The platform can be thought of as a computational implementation of the (model of) the real world system under study.

Initial runs of the platform are used to perform Calibration (163). This is needed to determine how to *translate* Domain parameters and variables into their corresponding platform values (for example how to translate between real world time, and simulated time), and how to take simulation experiment raw output data and *analyse* it to enable *comparison* with domain results captured in the Domain Model. The amount of calibration effort needed depends on Research Context (119) and Simulation Purpose (121).

1.7.4 Domain and Simulation Experiments

Within the CoSMoS approach the concept of an experiment is present at two stages: *domain experiments*, performed on a real world system within the Domain, and *simulation experiments*, carried out on the Simulation Platform.

These are analogous to *in vivo*/*in vitro* and *in silico* experimentation respectively.

A common goal of a CoSMoS Simulation Project is to run Simulation Experiment (177)s on the Simulation Platform that enable us to build a Results Model that can be compared to the Domain Model and provide insight back into the real Domain of study. This might help in understanding the results of domain experiments, or help in the design of new domain experiments that can further illuminate the domain model. So the Simulation Platform must allow us to run appropriate Simulation Experiments that reflect the concepts of interest in the Domain.

Specifically, Simulation Experiments should allow us to select the appropriate model components and behaviours, control the initialisation of key parameters, and perform appropriate analyses via suitable statistics. These abilities are explicitly represented within the Platform Model in terms of the Simulation Experiment Model (152).

There is a danger that the Simulation Experiments are used to explore hypothesised behaviours that are infeasible to observe or measure in the Domain, producing incomparable, uncalibratable results. So it is sensible to ensure that the Simulation Experiments are related to possible domain experiments, captured in the Domain Experiment Model (135), to help ensure comparable results.

Domain Experiment Model

The Domain Model can be viewed as describing the behaviours present in the Domain that are expressed when probed via domain experiments. The Domain Experiment Model is the place to explicitly model these domain experiments, describing the experimental system present in the domain, identifying, for example, experimental procedures and protocols, variables and ranges, controls, measurables, data volumes, sample sizes and statistical tests.

The Domain Model provides the Domain concepts and behaviours, and is factored into three component submodels (Figure 1.3):

- the Domain Experiment Model (135) identifies the model parameters and how we manipulate them
- the Basic Domain Model (130) captures the (usually hypothesised) Domain micro-level concepts, including components, structures, mechanisms, behaviours and interactions
- the Domain Behaviours (135) model captures the domain macro-level emergent observed behaviours

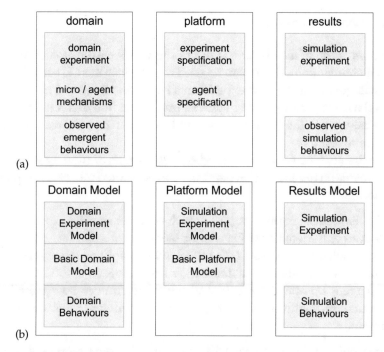

Fig. 1.3 (a) The substructure of the various CoSMoS components; (b) the mapping to the relevant CoSMoS patterns

Simulation Experiment Model

The Platform Model comprises computational representations of the relevant Domain Model components, making implementation abstractions, and is factored into two component submodels:

- the Simulation Experiment Model (152), which is derived from the Domain Experiment Model, and incorporates instrumentation for running experiments
- the Basic Platform Model (151), which captures the computational realisation of the Domain micro-level concepts

The deliberate lack of a component corresponding to the Domain Behaviours macro-level model helps ensure that the 'answer' – the emergent behaviours resulting from the hypothesised micro-level behaviours – is not explicitly coded into the Simulation Platform.

Simulation Experiments

Implementation of the Platform Model results in a Simulation Platform. A calibrated simulation platform can be used to run Simulation Experiments that are analogues of real world experiments run in the Domain. The results of a simulation experiment (after suitable translation into domain terms, and data analysis, via the Results Model) can be compared to the real world experimental results.

The Results Model comprises models of the results of running simulation experiments on the simulation platform, and is factored into two component submodels:

- the Simulation Experiment (177), which captures specific experiment instances of the Simulation Experiment Model
- the Simulation Behaviours (179) model, which captures the observed and analysed simulation behaviours in terms analogous to the macro-level emergent Domain Behaviours, and which determines how these behaviours are identified and measured when running Simulation Experiments

The results can be compared with analogous domain experiment outputs in the Domain Model. If they disagree, it may be because:

- the variables and parameters are not being translated appropriately (Calibration may have *overfit* their values)
- there a faults in the design of the Platform Model or in the implementation of the Simulation Platform
- there are faults in the Domain Experiment Model, such as imperfect measurements or statistical errors
- there are faults in the Basic Domain Model: the science is imperfectly captured, or imperfectly understood

1.7.5 Arguing fitness for purpose

A simulation is built for a purpose, and those using the Simulation Platform to run Simulation Experiments need to be confident that it is suitable for that intended purpose. The scientific results from using the Simulation Platform may be published, or may underpin further scientific research; in such cases, people outside the team developing and using the Simulation Platform need to be given reason to have confidence in its *fitness for purpose*.

To build the appropriate confidence in a particular simulation-based study, the team needs to present a rationale for the fitness for purpose of the entire simulation project (including modelling and simulator development, input data, and analysis of results).

The case for fitness for purpose comprises evidence of the quality of development, the consideration of assumptions, and the work on assuring the results. Rather than create a dense textual argument, a Structured Argument (189) can be created, with appropriate linkage to supporting evidence, assumptions and justifications. The fitness for purpose argument may be developed *post hoc*, or may be used to drive the shape of the simulation development process: it is easier to argue a system is fit for purpose if the development has been guided with such a need in mind.

We specifically use the terminology "fit for purpose" (with the meaning "good enough to do the job it was designed to do"[5]) and "appropriate". These terms emphasise that these properties are *relative*, to the Simulation Purpose, and hence that there is a need to revisit arguments should that purpose change.

We choose not to use more common terminology such as "valid" or "correct". These terms have implications of being absolute terms: "this instrument is correct", as opposed to "this instrument is *fit for a given purpose*". These stronger terms do not capture the need to revisit arguments if circumstances change. Additionally, they have implications of being either true or false: something is either "valid" or "invalid", whereas we want to capture a continuum of possibilities, allowing a Simulation Platform to have degrees of fitness for purpose.

A fitness-for-purpose argument is usually incomplete: its purpose is to capture the understanding about fitness for purpose of its audience, so that it can be referenced in future, challenged and revisited. A thorough and fully documented argumentation exercise is unnecessary in most situations, particularly in cases where the simulation criticality is low.

As well as documenting what is done, and arguing that it is the right thing to do, it is important also to document what is not done, and argue why it has been decided not to do it. This saves much grief later in the project, when a previously dismissed approach is retried, and the reason for its dismissal rediscovered.

It is important to note that, particularly in the context of complex systems simulation, an initial set of tests and calibration experiments may not fully exercise the Simulation Platform. Subsequent Simulation Experiments may later expose hidden assumptions, which are likely to clarify the scope and scale for which the simulation is appropriate (and outside which it cannot be "trusted"). Fitness for purpose should be viewed as a temporary state that may be contradicted by subsequent experimentation, either because the understanding on which the simulator was designed is inadequate, or be-cause the Simulation Purpose was not fully defined.

Chapter 2
What's in it for me?

Abstract — In which two Domain Scientists (Jim Bown and Alan Winfield) argue why the CoSMoS approach is beneficial to them.

2.1 Why this chapter, and why us?

CoSMoS is a sophisticated, powerful and detailed approach to support simulations, and it encompasses the whole process of conceptualising, designing, constructing, executing and evaluating simulations as scientific instruments. At the same time the CoSMoS approach is subtle and multi-faceted in supporting that process. This means that for the non-software engineer the CoSMoS approach may be perceived as on the one hand overwhelming and on the other hand abstruse. In this chapter, by taking a scientist's perspective rather than that of a software engineer, we hope to offer a valuable pathway into the comprehensive coverage of the CoSMoS approach provided in Part II.

We are well placed to do this since we are scientific project partners to the CoSMoS project, with backgrounds in systems biology [Bown] and systems engineering [Winfield]. Part of our role has been, in conjunction with other domain scientists associated with the CoSMoS project, to provide an external perspective on the role and value of much of the process developed here, together with a view on how that process may help address the issues that we currently face and are likely to face in the future.

We recognise that using the CoSMoS approach constitutes a significant investment of time on behalf of the domain scientist. The remainder of this chapter seeks to explain the relevance and value of the CoSMoS approach to scientists by highlighting what the CoSMoS approach may contribute to our

S. Stepney, F.A.C. Polack, *Engineering Simulations as Scientific Instruments: A Pattern Language*, https://doi.org/10.1007/978-3-030-01938-9_2

respective domains. We summarise briefly those domains, and indicate how the CoSMoS approach helps address some of the cross-cutting challenges in studying complex systems that are shared in biology and engineering. We also consider indicative challenges unique to each of our domains and likewise discuss the role of the CoSMoS approach in overcoming those challenges.

Our intention is to illustrate the value of CoSMoS for fields beyond our own. Many of the challenges we raise are manifest in other complex domains such as financial, environmental and socio-technical systems. Readers from other domains should find pathways into the value of CoSMoS by recognising some of the challenges we raise in their own arenas and in doing so form a lens through which to read Part II.

2.2 What challenges do we face

2.2.1 Systems biology

Biology is at the root of some of the most challenging and important problems facing society today, spanning diverse areas such as adequate food production, preservation of biodiversity, landscape management in the face of climate change, and maintenance of human health in a population increasing in age and size. Biological systems have for a long time been studied through reductionist science, decomposing the system into its constituent parts. While this has provided extensive data sets characterising the individual components, reductionist science has failed to provide a scheme for re-integration of that data into a holistic view of the system. As Cohen [54] states: "The more data we have access to, the more confused we have become."

Systems biology is a relatively new scientific discipline that seeks to provide that integrative view by treating biological systems as complex systems, i.e. a system that comprises many individual components that interact in space over time. Complex systems in biology are known to exhibit several phenomena that make them interesting but difficult to study, including:

- that they are inherently spatial, and so the spatial patterning of individual components impacts system dynamics and *vice versa*
- system components exhibit natural variation, and this variation is important to the interactions among these components

- component interactions generate emergent patterns, e.g. bird flocking and ant trails, visible at the system-scale but not deducible from the measurables of the individual components
- system dynamics are governed by processes operating at a range of spatial and temporal scales and this, combined with the difficulties associated with experimental measurement generally, means that no one experimental approach may characterise the system fully

2.2.2 Systems engineering

Almost all modern engineering is systems engineering. Even self-contained engineered artefacts like mobile phones or mobile robots are composed of many sub-systems (circuits or modules), which are systems-engineered to achieve the overall required functionality. The very high levels of functionality that are now commonplace in such artefacts owes itself to several critical factors: one is the standardization of modular sub-systems and their interfaces (think of a third-party designed and manufactured GPS receiver treated as a 'component' within a mobile phone or digital camera); another is the protocols that allow these sub-systems to communicate. The third is software, which embeds multiple layers of e.g. device drivers (often standard components), protocol stacks, and applications code; much of the functionality of engineered systems is achieved through software design and often that software provides the overall integration of sub-systems into a functional whole. Good software engineering is a profoundly important part of systems engineering.

Traditionally systems engineering has employed requirements-driven methodologies that proceed by first decomposing high-level requirements into successively lower level components, then – once that decomposition is complete – instantiating and testing those components (hardware and software), then successively integrating and testing until the overall system is complete. This is the so-called V-shaped model of systems engineering [182], although more recent approaches such as Agile Systems Engineering have challenged the V-shaped model [218].

Traditional approaches to systems engineering are not always successful, however, and failures of large-scale systems are well known [70]. While the reasons for such failures are often attributable to poor requirements or weak project management, less obvious factors can play a significant role, especially for large scale or distributed systems. These include: (i) an unpredictable or poorly understood and ill-specified operational environment, which might lead to (ii) unexpected emergent and/or stochastic behaviours from the system when operating in that environment, or (iii) unknown emergent behaviours arising from the combinatorial complexity of the large number of sub-systems and their interactions. And (iv) if failures of sub-systems, or simply component specifications drifting out-of-tolerance, occur then different kinds of unpredictable system properties will emerge.

Responses to these challenges to systems engineering include, for example, engineering the operational environment in addition to the system in an effort to address (i) and (ii); an example would be the engineering of a warehouse in order to provide autonomous guided vehicles for moving inventory with a fully characterised and predictable environment. In the case of large-scale VLSI designs the risk of (iii) is significantly minimized by employing advanced circuit simulations to exhaustively test the functionality of the circuit design before committing it to silicon. However, where engineering the environment is not an option as would be the case for unknown (natural or unexplored) environments, or known but unpredictable environments such as human living or workplaces, then modelling – of both the system and its environment – should, we propose, become part of a new systems engineering approach for complex systems.

2.2.3 Challenges in systems biology and systems engineering

These scientific domains are characterised by interactions among diverse individuals in space over both ecological and evolutionary time-scales. Complex systems models seek to simulate those interactions, which may be driven by processes at single or multiple spatio-temporal scales depending on the system of study. Additionally, system-scale patterns can emerge from individual-scale processes, and such non-linearities are characteristic features of many systems in engineering and especially biology. Based on this, we consider the following (non-exhaustive) set of fundamental challenges to modelling complex biological and engineered systems:

- the reality gap
- representing individuals and processes
- characterising emergent behaviour
- acceptance of model results

There are, of course, clear differences between biological and engineered systems. Engineering systems can be configured with a vast array of options and these options can impact significantly on system dynamics; in contrast biological system configuration is rather blunt with manipulation of bulk-scale parameters such as resource level or a small number of system-scale interventions. Engineered systems must build in mechanisms to be resilient in the face of failure and be explicit about their environmental operating range; in contrast resilience is a pervasive feature in many biological systems.

Engineering systems can be heavily instrumented and so it is possible to measure everything; in contrast, in many biological systems it is difficult to measure anything without disrupting the system under study and the often opaque physicality of the biology, e.g. soil systems or the human body, further impedes measurement. Finally, in both systems new knowledge is generated on a continuous basis, but this is particularly characteristic of biology with -omics high throughput technology and new instrumentation offering new spatio-temporal resolutions of study. As a result, we additionally consider the following domain-specific challenges:

- Simulation-based Internal Models – engineering
- A rapidly developing yet incomplete knowledge base – biology

2.3 What can the CoSMoS approach do to help?

CoSMoS provides a framework for developing complex systems models that makes systematic the identification of key elements that are required in such a model: the domain that establishes the system of study; the domain model that defines the purpose and scope of the simulation; the platform model that frames that domain model in an engineered design; a simulation model that encodes the platform in software and hardware; and a results model that enables interpretation of model output by domain experts. By making explicit the essential characteristics of each element and their interplay the task of defining a complex system becomes not only manageable but also illuminating. The explication of aspects of the system that might otherwise be implicit serves to challenge any assumptions surrounding those aspects and gives confidence when interpreting results from models based on those assumptions. The separation out of the different elements allows both a stepwise approach to model construction and reduces the likelihood of building the (assumed) answer into the model, a particular concern if a simulation platform is built directly from the domain with a view to obtaining specific (types of) results.

In order to illustrate the value of CoSMoS generally, we first consider the value of CoSMoS in the content of three domain-independent challenges that impact the domains of both systems engineering and systems biology. To show how CoSMoS can likewise support the nuances found in distinct problem domains, we then consider one characteristic domain-specific challenge in each of systems engineering and systems biology.

2.3.1 Domain-independent challenges

The reality gap

No simulation tool models a system with perfect fidelity. Like mathematical models simulators model an abstraction of the system; a simulator is a computational model of a system with limited fidelity. In robotics this loss of fidelity is known as the reality gap. We describe the reality gap here in the context of robotics and then biology, but it applies to all domains and is therefore a domain-independent challenge.

In the domain of robotics the use of simulation tools to prototype and test robot control algorithms is widespread and generally accepted as a valid way of speeding up the development process. Commercial robot simulation tools, such as Webots [162], semi-commercial simulators such as V-REP [195], or open source simulators such as Player-Stage [87], model both the robot(s) and their operational environment, and typically claim that code developed in the simulator can be transferred directly to real robots. All of these robot simulators have a pre-defined library of models for popular research robots, with the facility to add new robots (using, for instance, XML as a description language). However, there is also a well-known problem with robotics simulation: robot algorithms or code developed and tested in the simulator rarely works as expected when transferred to the real robot. The problem is especially acute when the simulator is used to artificially evolve the robot(s) controllers – because genetic algorithms tend to exploit simulation artefacts – but exists equally when robot controllers are hand-coded. This problem has become known as the *reality gap*: the gap between the simulated robot and its world, and the real robot in the real world [128]. But, although the problem is well-known, the presentation of new algorithms or results in robotics research, in which all development has been undertaken entirely in simulation and where there is little or no accompanying analysis of exactly how the results might be compromised by the reality gap, persists. When challenged, there is often a poor understanding of precisely where and how the results might have been affected by the simulation: for example, what are the effects of the 'idealised' models of robot sensors; or the effects of the particular approaches taken to model physical interactions by the 'physics engine' in the simulator; or why the lack of simulated environmental 'noise' might itself be a problem; or the effect of the simulator's discrete time steps.

The CoSMoS approach can help to address the problem of the reality gap, and the interpretation of what the reality gap means in robotics simulation, in several ways:

1. By drawing attention to the simulator and its role in the developmental process, and hence raising awareness of the need to understand the limitations of the simulator and what those limitations mean to the specific engineering task in hand, rather than simply taking it for granted that the simulator provides an appropriate model of the system and its environment. This will improve the analysis and hence the quality of claims made for robotics algorithms developed and tested entirely in simulation.
2. By allowing us to approach the measurement and calibration of a robotics simulation tool in a principled way, thus allowing us to analyse quantitatively, as well as qualitatively, the effects of the reality gap.
3. By then, again in a principled way, enhancing the simulated model. Thus, if the analysis of point (2) shows that the way a particular robot sensor is modelled has a significant impact on the overall quality of simulation, whereas the effect of all other abstractions is marginal, then by replacing the model of that sensor with a higher fidelity model (perhaps calibrated from measurements on real world sensors), we can significantly reduce the reality gap for the engineering task in hand. The same would be equally true for targeted enhancements to the environment model within the simulator. Here we are not only understanding, but reducing, the reality gap in a principled way.
4. By iterating on point (3), so that the robotics simulation, and the models within it, are co-developed alongside the robotics systems the simulator is being used to develop. This represents the full application of the CoSMoS approach, with significant benefits to the quality of the claims that can be made for the algorithms developed in simulation, and to the real world robotics systems ultimately developed.

In biology the reality gap is likewise present. Systems biology seeks to understand the relationship between measurable processes – often at the component scale – and observable patterns – typically at the system scale. To gain a thorough understanding of these processes often requires a level of control over system input that is not possible in realistic contexts and so a simplified system is studied. In many cases the elements within a complex system exhibit context-dependent behaviour and so results may not translate from the simplified system back to the real system. In CoSMoS terms, this desire to understand key processes can lead to a disconnect between the Domain and Domain Model and the difficulties of Results Model interpretation this disconnect then attracts. Here, we illustrate the reality gap with reference to cancer systems biology, although this challenge is certainly not limited to this system.

In cancer systems biology the overarching goal is to design effective anti-cancer therapy by studying the impact of candidate anti-cancer drugs on cell function and fate, thereby elucidating the processes by which drugs act on cells. Early stage exploration of drug compound action on cells is carried out using an experimental system comprising a monolayer of cells [124]. These 2D systems benefit from relatively high throughput, established experimental protocols and well-developed approaches to measurement of drug action over time. In spite of these benefits, there is very limited success in translating findings from these systems into clinical practice [80]. Part of the reason for this lack of translation is the complexity of the human cell. In addition to a cell's internal complexity (and see the next domain-independent challenge of emergent behaviour) a cell's behaviour is context specific. There is a growing body of evidence to show that cells behave differently in monolayer than in 3D structures such as the human body [191]. This is in part because of differences in nutrient availability and oxygen levels [139]: in 2D systems cells are uniformly exposed to nutrients and oxygen; in 3D systems there exist spatial gradients and in the case of tumours where cell packing is often denser than normal tissue these gradients can be acute. Likewise, drug application in monolayers approaches uniform treatment of cells; in 3D systems there are again spatial gradients [139]. Finally, the physical pressures introduced by spatial packing, which are largely absent in monolayer systems, introduce the phenomenon of mechano-transduction in cells – the triggering of internal mechanisms based on external forces that can drive tumour progression *in vivo* [38].

In response to these differences, new experimental systems have been devised, with the spheroid – an aggregate of cancer cells – emerging as a key model system. These 3D systems provide some of the properties of *in vivo* tissue: spatial architectures of cells; nutrient and oxygen gradients, mechanical stresses, etc. but are less standard in their protocols and more challenging to analyse [139]. Efforts have begun to characterise the differences between monolayer cell cultures and spheroids by carrying out experiments on those 2D and 3D systems using comparable conditions. Results show marked differences in behaviour exhibited by some cell types in 2D compared with 3D in response to some therapeutic insults (see for example [124, 191]), both in terms of key internal proteins and cell lifecycle dynamics. Additionally, there are a range of 3D systems available such as those based on bioreactor or bioprinting technologies, and each have their advantages and disadvantages with reference to the in vivo system they seek to inform: for example bioprinted allow high levels of control over physical architecture but seeding the systems with cells in a controlled pattern is challenging (see [194] for a review). Likewise, and following on from more than a decade of computational modelling of cells using data derived from monolayer experiments, computational modelling has begun to model 3D systems that are able to take account of the spatio-temporal development of tumours and response to treatment (see [133] for a review). Thus the field of cancer systems biology must on the one hand build on decades of drug-monolayer system research and on the other hand incorporate new findings from these new and emerging 3D systems.

The CoSMoS approach can help to address the challenge of the reality gap in systems biology as follows:

1. By guiding the development of new computational models of 3D systems that are designed to help understand tumour response to treatment by setting out the explicitly the relation between the Domain, 3D *in vivo*, the Domain Model, 3D *in vitro*, and the Platform Model, 3D *in silico*. There is a range of different *in vitro* 3D systems, each differently representing key aspects of *in vivo* systems to greater or lesser degrees. This range is also rapidly advancing in realism and sophistication. By establishing the relation between the Domainand Domain Model, CoSMoS provides a rigorous yet accessible way of capturing the inherent strengths and weaknesses, in other words what *in vivo* features are preserved and to what degree *in vitro*. Moreover, the explicit mapping of Domain Model to Platform Model, and then to the Simulation Platform, allows a clear linkage between the *in vivo* system and the *in silico* model, which then helps stakeholders interpret and ultimately accept model results as useful (see the subsection *Acceptance of model results* below).

2. By informing the integration of 3D systems data into new models, through the explication of Domain Model and Platform Model. Computational models of 2D cell systems concern themselves with time but not typically space. For example genomic and proteomic dynamics are averaged in space and measured over time at the whole system scale. 3D *in vitro* systems generate more complex, spatially structured data, and approaches for model integration are less well developed than for their 2D counterparts. 3D system data integration depends on statistical assumptions; CoSMoS is ideally placed to capture those assumptions at the interface between Domain Model, which can encapsulate the spatio-temporal data derived from the system, and Platform Model, which can

detail the statistical representation of that data under given assumptions.

3. By enhancing model interpretation through the Results Model and its corresponding Domain Model. The results from computational models of 3D systems are complex, and large data sets must be distilled down into the essential system readouts, a small subset of the total data available. CoSMoS can guide this reduction process by framing the results in the domain, ensuring that any model fitting – user driven or algorithmic – is guided by those essential system readouts.

4. By enabling the introduction of multi-stream data, in order to draw on the wealth of 2D data streams available. First, CoSMoS can make clear both the source of that 2D data in its Domain and the potential limitations of its relevance in 3D systems in the Domain Model. Then, it is possible to use CoSMoS to help bridge the gap between 2D and 3D data streams through principled inclusion of 2D data under different assumptions (Platform Model) with support for model fitting as in point (3).

Characterising emergent behaviour

It is of great importance that emergent properties of a real system also manifest themselves in a simulation of that system. In systems engineering emergent properties are of great interest because they are often undesirable, and point to design or implementation flaws. However, this is not always the case. In both systems biology and systems engineering we could be modelling systems of multiple interacting elements (cells or agents in a distributed system, for example), in which the desired systems behaviours emerge, perhaps from self-organisation of those cells or agents. In systems biology, for example, emergent properties often hold the key to linking processes to patterns.

In either case careful attention to emergent behaviour is needed, for several reasons. The first is that emergent behaviours apparent in a simulation might not in fact be present at all in the real system; instead they can arise because of so-called simulation artefacts. Such artefacts could be because of errors in the simulation model, including in physics modelling (which, if present, is typically provided by a plug-in module in many simulation tools), or they could be because of the simulated timing of simulated interactions (because simulators generally serialise what are in reality parallel processes). The second reason for caution in interpreting apparently emergent behaviours in simulated systems is that emergent properties can be brittle. A small change in parameters, including noise, can radically change whether emergent behaviours manifest or not. These factors mean that it is very difficult to make confident (qualitative and quantitative) claims about emergent behaviours in real world systems based on observations of those behaviours in simulation.

Continuing with the systems biology example above, the characterisation of emergent behaviour is of paramount importance in anti-cancer drug design. Contemporary drug design is based on a paradigm of a single drug for a single target in the cellular signalling network [158], where that drug seeks to restore (or at least limit) aberrant functioning at that target. A recognised mechanism of drug resistance is feedback loops, pervasive in biology and often providing positive regulatory effects, that in cancer cells can allow those cells to adapt to the effect of the drug [214]. These feedback loops emerge from the interaction of different components within a topologically complex intracellular signalling networks, characterised by subnetwork cross-talk and cross-activation together with built-in redundancy to preserve functioning in a wide range of circumstances [159]. Combination therapies seek to overcome those negative emergent behaviours by targeting multiple points in the network [52]. However, given network sensitivity to both the order and timing of drugs within a combination regimen, effective therapy design is very challenging indeed. Computational models can support rapid and cheap exploration of combination therapy designs, but to be useful we need to be sure that emergent behaviours are a consequence of the modelled system and its dynamics and – importantly – are not built into the model itself. Through its transparent documentation of the translation from Platform Model to Simulation Platform CoSMoS helps avoid emergent behaviours arising from simulation artefacts.

Here a good understanding of emergence and its origins together with careful attention to the reality gap, as outlined above, help to avoid issues with both simulation and interpretation of emergent behaviours.

Acceptance of model results

Modern simulation tools are powerful but also dangerous. Dangerous because it is too easy to assume that they are telling us the truth. Especially beguiling is the renderer, which provides an animated visualisation of the simulated world and the agents in it. Often the renderer provides all kinds of sophisticated effects borrowed from video games, like shadows, lighting and reflections, which all serve to strengthen the illusion that what we are seeing is accurate. It is important to understand that the sophistication of the scene renderer is not an indicator of the fidelity of the simulated world and the agents in it. Results should only be accepted on the basis of a sound understanding of the Domain Model and its limitations, the Platform Model and its limitations, and the Results Model – all supported by sound arguments. A key part of CoSMoS is argumentation: the fitness-for-purpose of the models can be presented as an argument. This argument presents both the underpinning evidence for a model and also its limitations and uncertainties. In doing so, CoSMoS provides a framework to structure those arguments into the various concerns as per the CoSMoS approach.

Where models are used to inform real world decision making, this argumentation is of particular importance. In healthcare for example, systems biology models are typically used to direct pre-clinical experimental effort, and in doing so seek to save experimenter time, expensive consumables and even reduce animal experimentation. As data increases in volume, resolution and realism models grow more sophisticated, and there is increasing interest in models for personalised medicine [116].

For example, Patel et al. [175] construct a systems biology model of toxicology of a drug used to treat depression (citalopram). They construct a mechanistic model coupled to an *in vitro* system and calibrate the model to individual patients. They validate personalised model predictions against individual patient case histories, and show good agreement. They highlight the importance of explicitly describing the model assumptions and limitations to "build confidence in the model".

More broadly, Karolak et al. [133] likewise recognise the potential value of computational models for personalised medicine, and provide a useful summary of work in the area of oncology. However, in medicine in particular such models can be met with resistance, since, even when sophisticated, they grossly oversimplify a highly complex system and seek to make quantitative a knowledge base that is often framed in qualitative terms [221]. We believe CoSMoS has a useful role in promoting useful model development in this challenging arena where confidence in model behaviour is paramount.

2.3.2 Domain-specific challenges

Engineering domain: simulation-based internal models

In recent years simulation has emerged as an embedded component within a new class of cognitive robotic systems, thus raising new possibilities and challenges. An embedded simulator provides a robot with a mechanism for generating and testing what-if hypotheses:

1. What if I carry out action x?
2. Of several possible next actions x_i, which should I choose?

This leads to the idea of a simulation-based internal model as a *consequence engine* – a mechanism for estimating and hence anticipating the consequences of actions [233].

Recent work has proposed and experimentally tested simulation-based internal models to provide robots with simple ethical behaviours [220, 232], improve robot safety [32] and to enable robots to infer the goals of another robot – so called *rational imitation* [219]. In swarm robotics the same approach has been used to explore exogenous fault detection [163] and O'Dowd et al. [171] demonstrate a simulator running within an embedded genetic algorithm to artificially evolve swarm behaviours.

In all of this work a simulator for both the robot and its environment, including other dynamic actors (other robots, in some cases acting as proxy humans), is embedded within the robot; it thus has a simulation of itself inside itself (an internal model). Typically that simulation runs in real time. Periodically, perhaps once every second, the simulation is initialised with the current state of the robot (position and pose, etc) and the world including other robots, then run – for a given number of simulated seconds into the future (typically 10 s) – for each of the robot's several next possible actions. For each of those runs the consequences of that action (to both the robot and other dynamic actors) are evaluated; then, based for example on a safety or ethical rule, the robot's next real action is chosen. For the embedded simulator some of these works [32, 232] make use of a modified version of Stage [87], while the rest employ a purpose-built simulator; all are relatively low fidelity[1] simulators with light-weight 2D models for kinetics and the physics of collisions.

In that work simulation is not used as a scientific instrument, but instead as an embedded component within an engineered system (although the purpose of these systems is to experimentally test research questions in robotics science). The use of simulation as an embedded component within engineered system does raise several new challenges. First is the question of how we find the right compromise between simulation fidelity and computational budget in a real time system. The second is about timing. When, and how often, does the robot need to initiate the process of internally simulating its next possible actions? And how far into the future should we simulate (and could we adapt that time according to the demands of the situation)? The third question is about validation and verification. This is especially problematic when we consider that an advantage of the simulation-based internal modelling approach is that, in principle, it can cope with unknown situations and environments, since the robot initialises the consequence engine with the situation in which it finds itself. For a deeper analysis of these questions, see [233].

Simulation-based internal models are at present the subject of research, but should they find real world application these questions, especially that of validation and verification, will need to be addressed. Embodied Simulation (239) captures some of these issues. We are confident that the CoSMoS approach will be of benefit to that verification and validation process, and so support adoption of robots able to anticipate the consequences of their actions.

Systems biology domain: a rapidly developing yet incomplete knowledge base

Knowledge in systems biology is both provisional and incomplete. As with any scientific endeavour, new experiments are designed to investigate (biological) mechanisms in novel ways to produce new data that characterises the system of study. These new data then require incorporation into any model of that system. Biology faces an additional opportunity with respect to measurement, and challenge with respect to modelling of large omics. Large omics data can be derived from multiple organisational levels of a system, such as genomic, transcriptomic and proteomic. These omics data should provide the key to a deep understanding of biological mechanisms, and ultimately personalised medicine, yet the systematic integration of omics data into computational models is challenging [8].

Beyond the scale of the data, the most difficult task ahead is to understand the connections within and among these data streams. At its most abstract level, this task becomes one of linking genotype to phenotype. This linking requires consideration of multiple omic streams from the genome through the epigenome, transcriptome, proteome and metabolome in order to understand the phenome [192]. These data streams are connected biologically, but those connections are not identified by the data streams alone. They must be inferred from a mix of systematic perturbation experiments (for example, Casado et al. [47] explore the link between gene mutations and cancer-inducing aberrations in proteins influencing cell survival and proliferation) and, importantly here, assumptions of interconnections among levels.

Even within a single level of study, system knowledge, and indeed system representation, is incomplete. Models focus on specific signalling pathways linked to specific function. However, any given component of that pathway might itself be influenced, directly or indirectly, by another level of organisation and by other pathways outwith the Domain of the system considered [144]. Moreover, any given experimental design cannot practically populate all components in the network, and so wider literature is drawn on for some measurements; model fitting is used to characterise those system components not measured directly or derived from literature. Accordingly, modelling must be carried out with incomplete knowledge.

Cvijovic et al. [58] provide a useful review of the challenges presented by gaps in knowledge and recommend, among other developments, the inclusion of stochastic descriptions of uncertainty in input data, constraints on parameter values informed by incorporating previous work, and processes for model reduction, for example through sensitivity analysis. These inclusions can introduce marked changes in model formulation and performance and should be carefully documented for transparency.

To support this, CoSMoS offers argumentation (as above) in general, and goal structured notation (GSN) in particular. GSN offers a scheme to explicate assumptions for all aspects of CoSMoS from Domain through to Results Model, including simplifications and omissions, in a rigorous manner. This scheme formalises the link between assumptions and the knowledge base, promoting transparency in both those parts of the model that are supported by evidence, and those that are hypothesis-based. Both evidence- and hypothesis-based assumptions are, of course, entirely acceptable in modelling, but it is crucial to distinguish one from another. As the knowledge base evolves, GSN may also provide a way of systematically introducing, and recording the introduction of, new knowledge into the model. In turn, GSN also supports identification of the ramifications of new knowledge on model functioning through this explication.

2.4 Summary and future perspectives

Closing the loop

Systems modelling is most powerful when there is a virtuous circle of real world experiments driving computational model development, and the analysis of the results of that model in turn informs experimental design to generate new data ... to then influence model development and so on [140]. Effective implementation of this virtuous circle is difficult, however. At its simplest, this loop can mean results from new experiments can mean new knowledge to be added into a model to improve predictions. Beyond this, however, new results might lead to model reformulation because of new knowledge about how components interrelate, changes to the scope of the system under study, or even fundamental changes to the way in which a system is considered, provoking entirely new question sets. Thus, closing the loop might require systems model development in many different ways.

CoSMoS unpacks the modelling process into distinct aspects that support such development. New knowledge can be represented in the Domain Model. Model reformulation is then explicitly propagated to the Platform Model and Simulation Platform where necessary. New questions and changes in scope can be propagated through the CoSMoS approach from Domainthrough to Results Model. In this way, CoSMoS offers a set of stages and considerations through which to manage effectively and transparently the virtuous circle of coupled experimental and theoretical systems.

Open Science

Open Science is the emerging practice of making some or all of a science project available for anyone to witness as the project progresses, typically making use of project web pages as a window into the project. There is, as yet, no widely accepted standard approach for open science. Open science encompasses a spectrum of activities, some of which are already widely practised such as deposition of papers in publicly accessible repositories; publication in open access journals; inclusion of datasets with published papers, or the maintenance of project websites. But the fullest expression of open science makes "everything – data, scientific opinions, questions, ideas, folk knowledge, workflows and everything else" available as it happens [169, p.32]. In what is referred to as Open Notebook Science "researchers post their laboratory notebooks on the Internet for public scrutiny [...] in as close to real time as possible" [203, p.S21].

In recent work we have argued that open science represents a new way of building and sustaining trust in science; that open science represents a new Trust Technology [96]. Given that CoSMoS addresses the issue of how we can trust simulation in science and engineering and argues, in §1.3, that we must go beyond open-source code and openly argue the fitness for purpose of a simulation in modelling a particular problem and domain. Open science, and especially open notebook science, provides a framework that complements the CoSMoS approach. Indeed the CoSMoS approach lends itself to open science, since the argumentation, which supports the models, provides a narrative that should be published alongside those models, as the project proceeds. Opening the argumentation for criticism, debate and refinement can only help to build confidence and trust in the fitness for purpose of the CoSMoS models and the insights they provide.

Beyond Systems Engineering and Systems Biology

In this chapter we have set out a consideration of the value of CoSMoS in systems engineering and systems biology. Complex systems modelling has value in other domains spanning the natural and social sciences, including environmental management, individual and societal behaviour, financial systems, and beyond.

Researchers in the natural sciences and engineering disciplines will find many overlaps with our accounts of complex systems modelling challenges. In the social sciences there is growing interest in the use of complexity science to understand emergent phenomena and stimulate new thinking on existing problems. For example, van Wietmarschen et al. [229] consider the opportunities that complex systems models of healthcare that moves beyond biological to integrate psychological and societal aspects are "essential for aligning and reconnecting the many institutions and disciplines involved in the health care sector". Spaiser et al. [202] provide an account of complex systems modelling in education, where they explore the social phenomenon of segregation in secondary schools and using a data-driven approach determine characteristic complex systems patterns such as tipping points.

The increasing interconnection of our socio-economic systems, locally and globally, means that complex systems models are of growing importance. However, those systems are arguably even more complex than those considered in these chapters, meaning that the reality gap will be more profound, emergence more difficult to first interpret and then harness, and gaps in knowledge will be particularly prominent – models must rely on assumptions and constraints. We believe that CoSMoS has a prominent role to play in developing systems models to address some of society's most troublesome problem spaces.

Chapter 3
The CoSMoS approach in miniature

Abstract — In which we give an overview of the entire CoSMoS approach in miniature, by use of a small example development, tailored to illustrate the core CoSMoS components.

3.1 Introduction – a running example

In this chapter, we illustrate the CoSMoS approach with a running example. The example develops a simulator to support exploration of a particular theory of prostate cell division and differentiation [153], devised as part of a study of prostate cancer neogenesis that is expected to lead to novel therapeutic treatment of prostate cancer[1].

We present the example through instantiations of the relevant CoSMoS patterns. We use three distinctive styles to draw attention to the three kinds of exposition that comprise each instantiated pattern:

> **Context Pattern Name > This Pattern Name, or contextual step**
>
> > **This Pattern Name (page number of full description)**
> > Pattern intent and (optionally) summary, for navigation through the pattern structure.
>
> > (Optionally) further explanatory commentary, expanding intent and context of the CoSMoS approach.
>
> The content of the instantiated pattern for the running example.

© Springer Nature Switzerland AG 2018
S. Stepney, F.A.C. Polack, *Engineering Simulations as Scientific Instruments: A Pattern Language*, https://doi.org/10.1007/978-3-030-01938-9_3

The order of the boxed instantiated patterns below follows the structure of the CoSMoS patterns. This presentational order does not necessarily reflect the order in which the components were developed and completed: the development of some of the components proceeded in a different order, in parallel, and during overlapping phases. Furthermore, this is an incomplete presentation: some of the more-conventional material has been omitted, for space, and to help highlight the CoSMoS-specific aspects of the development process. Some details have been simplified or changed, for expository purposes.

The entire process is not needed for every project: CoSMoS can be tailored to a variety of different needs, see Partial Process (230). In particular, using a Multi-increment Simulation (240) approach would produce each model in an incremental manner. See Part IV for a larger example development.

3.2 The CoSMoS Simulation Project

> **CoSMoS Simulation Project (92)**
>
> ---
>
> **CoSMoS Simulation Project (92)**
> Develop a fit for purpose simulation of the complex scientific domain of interest.
>
> - carry out the Discovery Phase (95)
> - carry out the Development Phase (96)
> - carry out the Exploration Phase (97)
> - Argue Instrument Fit For Purpose (186)
>
> ---
>
> This top level pattern provides a route to build and use a simulation platform for scientific research, and argue it fit for purpose. Following the phases provides systematic development that aids scientific reproducibility, and supports subsequent modification and interpretation. A single increment comprising each phase should produce the set of models outlined in §1.7.2, with further increments modifying and extending these models.

3.3 Discovery phase

CoSMoS Simulation Project (92) > **Discovery Phase** (95)

Discovery Phase (95)
Decide what scientific instrument to build. Establish the scientific basis of the project: identify the domain of interest, model the domain, and shed light on scientific questions.

- identify the Research Context (119)
- define the Domain (123)
- construct the Domain Model (128)

3.3.1 Identify the research context

Discovery Phase (95) > **Research Context** (119)

Research Context (119)
Identify the overall scientific context and scope of the simulation-based research being conducted.

- provide a brief *overview* of the research context
- document the *research goals* and project scope
- agree the Simulation Purpose (121), including criticality and impact
- identify the team members and their experience, and assign Roles (99)
- Document Assumptions (108) relevant to the research context
- note the available *resources*, timescales, and other constraints
- design and set up a Project Repository (219)
- determine *success criteria*
- decide whether to proceed, or walk away

The Research Context identifies the overall scientific context and scope of the simulation-based research being conducted. It is the place to collate and track any contextual underpinnings of the simulation-based research, including the scientific background, and the technical and human limitations (resources) of the work.

The scientific context can be captured by recording high-level motivations or goals, research questions, assumptions, hypotheses, general definitions, and success criteria (how will you know the simulation has been successful). It is important to identify when, why and how these change throughout the course of developing and using the simulation. The scope of the research determines how the simulation results can be interpreted and applied.

Documenting assumptions on the research context helps to capture the general motivation and rationale for engaging in a collaborative simulation. This is one of the aspects of the research context that may be developed incrementally throughout the project, as new assumptions are uncovered.

Research Context (119) > overview

The overview provides context, and will be broader than the Simulation Purpose (121), which fits into this context.

Cancer is a phenotype arising as the result of aberrant interactions between many individual cells. The study of cancer is therefore the study of cell population dynamics. Mutations in a wide range of genes (*oncogenes*) are known to increase the risk of cancer [46]. Although the presence of mutations in oncogenes confers an elevated risk of cancer, there is a high level of variability in the timing and exact genotype of cancers. This stochastic element makes the analysis of cancers at the expression level very difficult. Stochasticity and genetic variability make cell population modelling a very attractive tool for the study of cancer neogenesis.

Cancer Research UK provides various statistics on prostate cancer incidence [45], including: in males in the UK, prostate cancer is the most common cancer; it accounts for 26% of all new cancer cases in males in the UK (2015); 1 in 8 men will be diagnosed with prostate cancer during their lifetime.

Research Context (119) > **research goals**

> We use the **!! Future** tag to mark aspects that will be important for future increments in a Multi-increment Simulation (240), but that are not part of the current increment.

The overall goals of the research are:

1. to create a dynamic, cell-based model of normal prostate epithelium that captures the processes of cell division and differentiation
2. **!! Future:** to augment this model in order to incorporate pseudo-genome inheritance
3. **!! Future:** to calibrate this model using primary prostate sample data
4. **!! Future:** to perturb the calibrated model by introducing rare (stochastic) mutations to individual cell states, in order to investigate the emergence of a cancerous phenotype from a dynamic population of differentiating cells

Only the first of these goals is addressed in the running example here. The subsequent goals are to be realised using Multi-increment Simulation (240). However, these subsequent goals do influence the design of the initial increment.

This first increment has the form of a feasibility study, to determine if the research programme is feasible. Specifically, the work should determine suitable modelling and implementation approaches, determine the computational resources needed for a full scale simulation, and bring the development team up to speed with the Domain (123).

Research Context (119) > **Simulation Purpose** (121)

Simulation Purpose (121)
Agree the purpose for which the simulation is being built and used, within the research context.

- define the *role of the simulation*
- determine the *criticality of the simulation results*

Identifying and agreeing the Simulation Purpose is key to shaping the of the rest of the process. The simulation has a purpose, a role to play, within the overall research context. Without a defined purpose, it is impossible to scope the research context, and it is impossible to arrive at a consensus over fitness for purpose.

The Simulation Purpose may, however, evolve over time, particularly as knowledge of the domain, and implementation possibilities and constraints, become clearer as the project progresses. Any changes to the purpose, and knock-on effects, need to be documented.

Simulation Purpose (121) > **role of simulation**

The overall aim of this project is:

1. Develop a model and simulation of prostate cell differentiation and division, based on prostate cell populations from laboratory research data. The simulation should replicate observed cell population dynamics, represented as changing proportions of cells in a "normal" prostate.
2. **!! Future:** Building on the model of the "normal" prostate, develop simulations that capture known environmental variation and mutation, in order to explore the emergence of cell proportions indicative of cancer (or other prostate conditions).
3. **!! Future:** Using these models of normal and cancerous prostate cell behaviours, develop simulation experiments that can be used to guide and test laboratory hypotheses of cancer development and control.

This first increment is an initial feasibility study, to demonstrate whether the simulation approach is computationally feasible. So the purpose of this increment is:

- to develop a model and simulation of prostate cell differentiation and division, where

   1. the Domain Model is agreed suitable by the Domain Scientist
   2. the simulation replicates observed cell population dynamics, represented as changing proportions of cells in a "normal" prostate
   3. the simulation can be run at a sufficient scale to investigate low probability events (mutations of single cells)

Simulation Purpose (121) > criticality of results

As a feasibility study, the specific simulation results are non-critical, and so lightweight arguments of fitness for purpose are all that is needed for these. As the purpose of this phase is to decide whether to proceed with a full simulation, argumentation related to scaling properties and resource requirements is more critical.

Subsequent phases will almost certainly be of higher criticality; the main risk is of publishing results from simulation experiments without good scientific evidence to support them, resulting in potential waste of resources by basing experiments on those results. So later phases should consider any developments in this initial phase as a Prototype (213), to be used to inform subsequent phases, but not to be used as the basis of a Multi-increment Simulation (240).

Research Context (119) > Roles (99)

Roles (99)
Assign team members to key roles in the simulation project.

- identify the Domain Scientist (101)
- identify the Domain Modeller (103)
- identify the Simulation Engineer (105)
- identify the Argument Modeller (106)
- identify other *optional roles*
- identify necessary *collaborations* between roles

The team comprises the following:

- the Domain Scientist: Norman J. Maitland, a senior academic cancer specialist, assisted by his team of laboratory scientists including Fiona Frame and other post-doctoral researchers, all expert in the domain of prostate cancer, but with limited simulation experience
- the Domain Modeller: Alastair Droop, a senior post-doctoral researcher, with knowledge of the domain, and with some modelling and implementation experience, assisted by Susan Stepney and Fiona Polack, two modelling experts and academic computer scientists
- the Simulation Engineer: Philip Garnett, a junior post-doctoral researcher, with expertise in developing biological simulations using the CoSMoS approach, but no initial knowledge of the specific domain
- the Argument Modeller: Fiona Polack, an academic computer scientist, with expertise in developing fitness-for-purpose arguments, but no initial knowledge of the specific domain

A Scribe was nominated at each meeting, to make notes of the discussions and decisions.

Intense collaboration was needed between the Domain Scientist and Domain Modeller, to communicate the required knowledge. Multiple meetings were held of all the team members, discussing the Domain, and validating the Domain Model; cake was a prominent feature.

Research Context (119) > Document Assumptions (108)

Document Assumptions (108)

Ensure assumptions are explicit and justified, and their consequences are understood.

- identify that an assumption has been made, and record it in an appropriate way
- for each assumption, determine its nature and criticality
- for each assumption, document the reason it has been made
- for each reason, document its justification, or flag it as "unjustified" or "unjustifiable"
- for each assumption, document its connotations and consequences
- for each critical assumption, determine the connotations for the scope and fitness-for-purpose of the simulation
- for each critical assumption, achieve consensus on the appropriateness of the assumption, and reflect this in fitness for purpose arguments
- revisit the Research Context (119) in light of the assumption, as appropriate

---

In documenting assumptions, we are seeking to record assumptions that fundamentally affect our understanding or modelling. It is impossible to document all assumptions, and many assumptions turn out to be irrelevant to the fitness of a simulation exercise. In general, however, it is good practice to record any recognised assumptions.

Documented assumptions provide input to the argumentation phase. They also provide a way of interrogating and checking modelling and other decisions. Documenting justifications helps to expose unneeded or unjustified assumptions. Documenting consequences helps ensure the the Simulation Purpose (121) is still achievable.

Assumptions are never "complete". Assumptions may be identified at any point in a simulation development. Late identification of an assumption and its consequences may require you to Propagate Changes (168) through the various models.

Here the Research Context (119) assumptions are documented, but the reasons, justifications, and consequences are mostly omitted. See later (Domain assumptions) for a more fully documented list.

A.1 Simulation will help provide insight.

> **reason:** Human data is limited, with small numbers of samples from each subject (biopsies are not usually performed regularly or repeatedly).
>
> **reason:** Animal models, which traditionally provide an experimental surrogate for human conditions, are not considered particularly helpful in this context.

A.2 Observation of the embryonic development of the prostate provides good insight into cell dynamics.

A.3 The proposed model of cancer neogenesis can be sufficiently expressed by a model of the amplification of rare mutation by natural cell division and differentiation.

A.4 The context is simulation of a human prostate

> **consequence:** results will not be directly applicable to other organs or other species

Research Context (119) > resources, timescales, and other constraints

- The Domain Modeller and the Simulation Engineer are each employed half time on the project, so each have two months effort. The domain scientist and his team, and the modelling and argumentation experts, can each provide ad hoc input of up to a few days effort each.
- The work has access to a local small compute cluster, for running simulations and gathering performance metrics.
- The timescale of the work is four months.
- There are no specific publication format constraints.

Research Context (119) > Project Repository (219)

> **Project Repository (219)** : Use a project-wide repository to co-ordinate the project information.

We use a project wiki for storing documentation and notes, and a GitHub repository for code.

Research Context (119) > success criteria

This initial increment is a feasibility study, designed to answer the following questions:

1. can we build a model that is both biologically plausible (provides a scientifically sound Domain Model) and is simulatable (provides an implementable Platform Model)?

   a. do we have adequate Domain knowledge?
   b. do we have adequate modelling approaches?

2. do we have adequate biological (laboratory) data for Calibration and to initialise the Simulation Platform?
3. can we use the Simulation Platform to run Simulation Experiments of a sufficiently large number (millions) of cells in a reasonable time (that run in less than a day)?

The answers to these questions will help determine if later increments are to go ahead.

Research Context (119) > decide

The decision was made to continue with the CoSMoS Simulation Project (92).

3.3.2 Define the domain

Discovery Phase (95) $>$ **Domain** (123)

> **Domain** (123)
> Identify the subject of simulation: the real world biological system, and the relevant information known about it.
>
> - draw an explanatory Cartoon (124) of the domain
> - provide an overview description of the domain
> - define the Expected Behaviours (126)
> - provide a Glossary (127) of relevant domain-specific terminology
> - Document Assumptions (108) relevant to the domain
> - define the scope and boundary of the domain – what is inside and what is outside – from the Research Context (119)
> - identify relevant sources: people, literature, data, models, etc.

The Domain (123) is the Domain Scientist (101)'s view of the subject of simulation; for example, a real world system that is the subject of scientific research, or an engineered system that is the subject of engineering research and design. This view might be a controversial view; it might be a view that the research is trying to disprove. It is the target for simulation. In contrast to the Research Context (119), which says *why* the simulation is being built, the Domain (123) describes *what* is being simulated.

The Domain (123) pattern provides information about the scientific domain, including references to source material and data, that the Domain Model (128) draws on.

The Domain is the process of cell division and differentiation in the prostate.

Domain (123) $>$ **Cartoon** (124)

Cartoon (124) : Sketch an informal overview picture.

An explanatory Cartoon helps to show the relevant components.

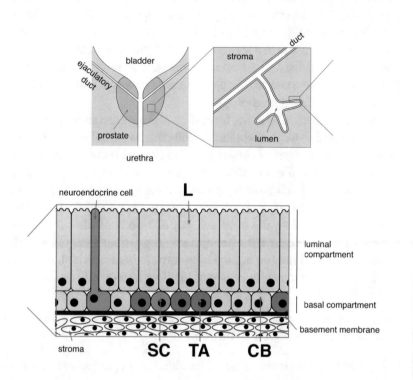

Fig. 3.1 A Cartoon of the modelled prostate cell types, showing the context within the gland, and identifying the four types of modelled epithelial cells: stem cells (SC), transit amplifying cells (TA), committed basal cells (CB), and luminal cells (L)

Domain (123) > **overview**

Several cell populations are present in the prostate (see Cartoon in fig 3.1). The majority of prostatic tissue is composed of stromal cells which consist of connective tissues and blood vessels. The tissue of interest with respect to prostate cancer is the prostatic epithelium, which consists of basal, secretory and neuroendocrine cells [55]. The secretory cell population consists of terminally-differentiated columnar cells. The basal cell compartment contains less-differentiated cells that are still in contact with the basement membrane.

The presence of small numbers of self-renewing stem cells in most tissues is now widely accepted. The stem cell population is able to replace dead cells in the tissue by the processes of division and differentiation. Stem cells are able to undergo two types of division: symmetric division in which a single stem cell divides to yield two similar daughter stem cells; and asymmetric division in which a single stem cell divides to give rise to a daughter stem cell and a daughter cell of a more differentiated phenotype.

The role of stem cells in the formation and maintenance of tumours is not well understood. Viable mutations in a stem cell will be passed on to its large number of progeny by the normal processes of cell division and differentiation. The cancer stem cell model [153] suggests that, if these stem cell mutations confer a malignant phenotype, then these progeny will form a cancerous population (a tumour). If this model is correct, then the stem cell population is of utmost importance in cancer treatment; the common therapy of removal of the bulk tumour mass will not impact the long-term patient survival, whereas ablation of cancerous stem cells would allow successful treatment.

Domain (123) > **Expected Behaviours** (126)

Expected Behaviours (126) : Describe the hypothesised behaviours and mechanisms.

This description encapsulates a summary of what is observable in the Domain, what concepts we believe to be involved and how, and what hypotheses related to these observables we want to investigate.

The expected observable is the experimentally-determined typical cell ratios, and the change in time from "normal" to "abnormal" ratios, indicative of a particular prostate condition such as cancer.

!! **Future:** abnormal cells to be modelled in later increments.

Figure 3.2 shows the system level transitions, and examples of cell states. All cells may have internal state transitions, but only two are shown: we know that there are potentially more relevant cell-level details. Note that Luminal cells are "active" in terms of division and differentiation when they are not active in terms of their secretory role in the prostate.

Three hypotheses related to cell dynamics to explore are:

1. Stem cells divide and differentiate successively to transit amplifying, committed basal and luminal cells; cells can maintain a steady state (dynamic equilibrium) through the division and differentiation plus cell death.
2. The move over time of the system to abnormal cell ratios (indicative of cancer or other prostate abnormalities) arises from small changes in the transition dynamics as expressed in (1).
3. Small changes in cell dynamics arise from mutable and heritable states within individual cells, for instance, caused by system level (intercellular) effects such as crowding, cell level effects (local mutation), and extracellular effects (due to environmental stimuli, where the mutation effect is important to the model, but the sources of the stimuli are not).

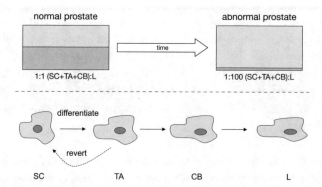

Fig. 3.2 Expected Behaviours figure for the case study. The expected observed behaviours are shown above the horizontal dashed line: the ratio of luminal to non-luminal cells in a normal prostate is 1:1 (more precisely, it lies in a range from 40:60 to 60:40); the ratio of luminal to non-luminal cells in an abnormal prostate is greatly increased. A cartoon of the lower level behaviours hypothesised to give rise to these is shown below the line. The four cell types are: **SC**, stem cell; **TA**, transit amplifying cell; **CB**, committed basal cell; **L**, luminal cell. Cells can differentiate or revert into other cell types, in the given sequence

Domain (123) > Glossary (127)

Glossary (127) : Provide a common terminology across the simulation project.

The main biological terms used in the various models are:

apoptosis – cell death
committed basal cell (CB) – one of the four modelled cell types
daughter cell – a modelling construct: a cell immediately after division, before differentiation
differentiation (diff) – a cell changes type
division (div) – one cell becomes two
epithelial cell – the four modelled cell types: SC, TA, B, and L
luminal cell (L) – one of the four modelled cell types
stem cell (SC) – one of the four modelled cell types
transit amplifying cell (TA) – one of the four modelled cell types

---

Domain (123) > **Document Assumptions** (108)

> When exploring the Domain, a wide variety of clarifications and assumptions are made. Domain assumptions are an opportunity for the Domain Scientist (101) to record their understanding of the limitations of the simulation activity, and an opportunity to record the answers when the Domain Modeller (103) and Simulation Engineer (105) Ask [Silly] Questions (217) while trying to understand the domain.

A.5 We can classify epithelial cells into four distinct categories: stem cells, transit amplifying cells, committed basal cells, and luminal cells.

   reason: In reality there is a continuum of cell differentiation; we focus on biologically distinct (stem and luminal) or distinguishable (committed basal and transit amplifying) cell types.
   justification: These four kinds of cells can be biologically distinguished due to their activity/secretions; the Domain Scientist can provide relevant cell counts or ratios for input to the simulator and for results comparison.

A.6 We consider only epithelial cells to be important in the development of prostate cancer.

   reason: Considering only epithelial cells will lead to a model that is simple enough to interpret against directly accessible biological research.

justification: The Domain Scientist notes that both blood vessel formation (via the stroma) and hormonal interactions (via neuroendocrine cells) are known to be important in prostate cancer development; however, the cancer role and mechanisms of the stromal and neuroendocrine cells is not well understood; the stroma and neuroendocrine effects are probably sufficiently distinct from the epithelial processes to be ignored (or captured as general environmental inputs).

consequence: The simulation will not be appropriate for study of any cancer, or other prostatic syndrome in which cell division and differentiation are implicated, that is influenced by, for instance, blood supply, or any condition in which stromal or neuroendocrine cells, as well as epithelial cells are implicated.

A.7 We ignore spatial aspects such as cell positions and cell contact.

reason: For model and implementation simplicity, and lack of sufficient biological data to model spatial aspects with any fidelity.

justification: It fits the research context: although spatial aspects, such as crowding and proximity, are thought to be important in, for instance, cell death and cell-to-cell signalling, extending the simulation to include spatial parameters is a level of complication that we do not wish to consider here; there are other ways to study spatial effect, and potentially a different simulator to be considered, which is beyond our resources in this study; additionally, the biological data for the prostate cell rates and ratios is aspatial.

consequence: We need to ensure that there are surrogates for relevant spatial effects (for example through relating cell death to cell numbers); we need to ensure that environmental influences include those that might, in reality, have an effect via spatial aspects.

A.8 We can express the cell division process as simple cell duplication.

reason: To simplify the model.

justification: In reality, a cell genome divides first, then the cell divides to accommodate each new genome; however, at the levels of abstraction used here, there is no functionality associated with the process of division; we are interested only in the abstraction to rates or probabilities of division and differentiation, and in mutation that affects cell division and differentiation rates.

consequence: Representing division as a single process means that we cannot add genome-level influences directly; cell division in the simulator needs to represent an appropriate abstraction of heritable genomic information, and needs to allow the possibility of faulty genome copying (mutation).

A.9 We can ignore the genes that control cell behaviour in response to extra-cellular environmental influences.

reason: The mechanisms of intra-cellular behaviour are better studied in the laboratory, or at lower levels of abstraction than we consider here.

justification: All environmental influences of relevance to our concerns lead to either the acceleration or deceleration of cell division and differentiation; the full set of environmental influences and the consequent intra-cellular behaviours are not known, and are not of relevance to this simulator.

consequence: We cannot use the simulator to explore different environmental influences, only the consequences of changes to cell division and differentiation rates.

Domain (123) > scope and boundary

- inter-cell-level modelling
- model cell division and differentiation
- consider only biologically-distinguishable cell states
- do not model blood flow to tissue
- do not model spatial aspects (assumption A.7)

Domain (123) > relevant sources

The main relevant sources are:

- the input of the Domain Scientist and his team of laboratory scientists, who provide
    - biological domain knowledge for modelling decisions
    - specific experimental data on division and differentiation rates, and cell number ratios

- the references given in the Research Context and the Domain overview

3.3.3 Construct the domain model

Discovery Phase (95) > **Domain Model** (128)

Domain Model (128)

Produce an explicit description of the relevant domain concepts.

- draw an explanatory Cartoon (124)
- discuss and choose the domain Modelling Approach (111) and level of abstraction
- define the Domain Behaviours (135)
- build the Basic Domain Model (130) using the chosen modelling approach
- build the Domain Experiment Model (135)
- build the Data Dictionary (132)
- build the domain Stochasticity Model (154)
- Document Assumptions (108) relevant to the domain model

The Domain Model explicitly captures understanding of the Domain (123), identifying and describing the components, structures, behaviours and interactions present in the Domain at a level of detail and abstraction suitable for addressing any identified research questions to be posed of the Simulation Platform (161). It is a model based on the science as presented by the Domain Scientist (101), and its design should be free from Simulation Platform implementation bias; it separates the model of the science from the implementation details of the Simulation Platform.

The Domain Model is a tool to exchange and discuss domain understanding between Domain Modellers and Domain Scientists. The reviewed and agreed Domain Model forms the agreed scientific basis for the eventual Simulation Platform.

Domain Model (128) > Cartoon (124)

> This Cartoon looks more formal than the Domain (123) Cartoon
> (Figure 3.1). However, it is still a Cartoon: there is no defined lan-
> guage, no semantics, no set of defined language concepts, just
> arbitrary boxes and arrows to help capture and explore under-
> standing.

An explanatory Cartoon shows the relevant processes of prostate
cell division and differentiation:

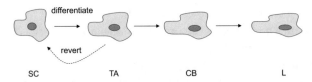

Fig. 3.3 A Cartoon capturing which cells differentiate, and which cell differenti-
ations might be reversions to a previous sort of cell: TA cells can revert to SC

During the discussion which led to drawing of Figure 3.3, it was
discovered that there is no agreed definitive understanding of the
process of reversion. One Simulation Experiment (177) that can be done
is to determine whether the agreed set of cell transitions in the model
are capable of capturing normal dynamics.

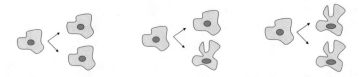

Fig. 3.4 Possible representations of cell division patterns: after division, each resulting cell is either the same type as the original cell, or is the next type in the
differentiation pathway

Domain Model (128) > Modelling Approach (111)

> **Modelling Approach (111)** : Choose an appropriate modelling
> approach and notation.

The Domain Model is the first stage in moving from Cartoon (124) to more formal software engineering models. The Domain Modeller (103), in collaboration with the Domain Scientist (101) and the rest of the team, identify the Modelling Approach.

In software engineering terms, the modelling language(s) used must be able to express the structure and interplay of system behaviours: it is not appropriate to model solely the data structures (class models and the like), or solely the structure of operations (sequence diagrams). With this proviso, it is essentially unimportant what engineering procedures, modelling and implementation languages are used, so long as it is possible:

- to express the Domain Model in a way that is clear to Domain Scientists and useful to the Domain Modeller and Simulation Engineer
- to trace how the components (structures, behaviours) of the Domain Model are captured in the implemented Simulation Platform (161)

It is also worth taking into account any presumptions about the kind of simulation that will be created, and the kind of experiments that might be carried out in simulation.

We use a form of Agent Based Modelling (208), because, in later increments, we want to be able to model mutation and heritability of individual cells (agents in the simulation), and to be able to track cell changes through the population.

The Domain Model need not include a notion of space (assumption A.7). However, we note that the prostate is a constrained space, and some division and differentiation is constrained by crowding; this will require us to introduce a crowding parameter to the simulation platform (assumption A.10).

We want a notation that clearly captures the processes of cell division (one agent becoming two agents) and of differentiation (one agent becoming a different type of agent). The Domain Scientist noted that Petri nets[2] might be suitable; Petri nets are attractive to many systems biologists, being widely used to model signalling and pathways in biological systems (see, for example [49, 93, 157, 197]). The Domain Modeller confirmed that Petri nets provide a clear and natural way to model division and differentiation, using tokens to model the cells, and transitions to model the processes of division and differentiation.

While building the Domain Model, we discovered that we also needed a notation that clearly captures the processes occurring in a cell during the part of its lifecycle where it is not undergoing the transformational changes of division and differentiation. Although Petri nets can be used to model such state transitions, the notation rapidly becomes unwieldy. We instead model the individual cells as state machines.

Because we are using two separate modelling notations, rather than creating a custom modification of one language, we have a Hybrid Model (212), and so we need to define how these different components fit together. At this domain modelling stage, it is sufficient to do so by defining naming conventions to link the models: the two layers are connected by the Petri net transition names and corresponding names that label the state diagram entry and exit points (see [63] for the full definition).

Domain Model (128) > Domain Behaviours (135)

> **Domain Behaviours (135)** : Describe the observed emergent behaviours of the underlying system.

The emergent domain behaviour of interest here is the cell number ratios (recorded in the Data Dictionary (132)). In the real domain, cell numbers change continuously, but the broad range of the ratios are reliably used as indicators of a normal or cancerous prostate (summarised in Figure 3.2).

In this first increment, we are interested in capturing the emergence of cell ratios in the "normal" range as indicated by the Domain Scientist.

Domain Model (128) > Basic Domain Model (130)

> **Basic Domain Model (130)** : Build a detailed model of the basic domain concepts, components and processes.

Generic model of cell division/differentiation

The various division/differentiations possibilities shown in the cartoons above (figs 3.4 and 3.3) can be captured in a single Petri net
model, shown in fig 3.5.

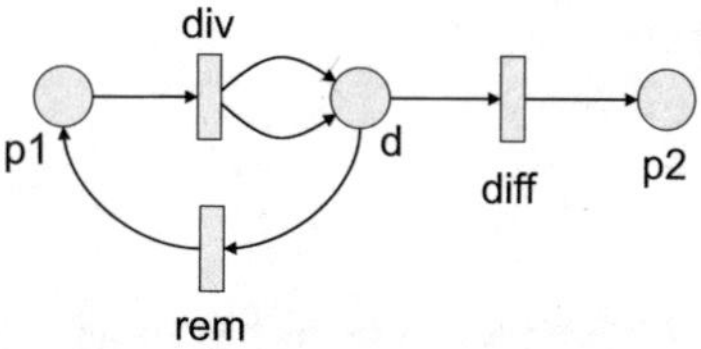

Fig. 3.5 An abstract Petri
net model of cell division
and differentiation: a cell
of type p1 divides into
two daughter cells, each
of which either remains of
type p1, or differentiates to
type p2

> This abstract model combining several concrete cases was de-
> veloped by the Domain Modeller, and then discussed with and
> agreed by the Domain Scientist (101) and the team during a re-
> view meeting. The abstract 'daughter cell' concept is a modelling
> concept, not present in the Domain or in any domain model Car-
> toon. So it is important to check whether this modelling concept
> is acceptable at a Domain level, or whether it should be deferred
> to a later modelling stage.

Model of stem cell division/differentiation

See Figure 3.6. A stem cell can divide (div_{SC}); its daughters D_{SC} can
each either remain a stem cell (rem_{SC}), or differentiate ($diff_{dSC}$) to a
Transit Amplifying cell TA. A stem cell can also differentiate ($diff_{SC}$)
directly to a TA cell without division, or can undergo apoptosis (ap_{SC}),
and become dead ($Dead_{SC}$). A TA cell can revert (rev_{TA}) to a stem cell.

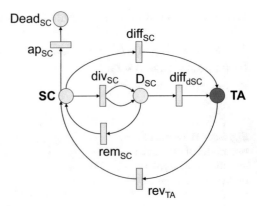

Fig. 3.6 A Petri net model
of stem cell (SC) division
and differentiation

The TA place is shown as a "fusion place", indicating that it occurs
in another Petri net model.

The models for the division and differentiation for the other cell
types are similar; see [63] for the full model.

Model of stem cell state transitions

See Figure 3.7. A stem cell SC has two sub-states: quiescent and active.
It is produced (via reversion, or from a divided daughter cell) in the
active state, and it exits to differentiate or divide from the active state.
It can apoptose from any state. The entry and exit points correspond
to the respective transitions in the Petri net.

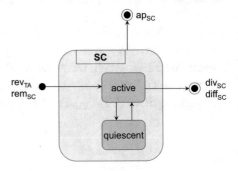

Fig. 3.7 A state machine
model of a stem cell SC

The state machine models for the other cell types are given in [63].

Transition rates

A final consideration is the control of the transitions in both the Petri nets and the state charts. The Domain Scientist can provide approximate cell ratios in each type, some indicative cell counts, and some suggestions of proportions of time cells spend in their various states.

We can convert the information on ratios and counts into probabilities, and use random numbers to determine whether a transition is available. At this stage, we give each cell a mutable, heritable pseudo-genome, which stores (at least) its transition probabilities. These probabilities could mutate during transition (Petri net transitions), due to intra-cellular factors (such as the overall-cell-count surrogate for crowding), or due to extra-cellular factors introduced when we seek to simulate cancer neogenesis.

Domain Model (128) > Domain Experiment Model (135)

> **Domain Experiment Model (135)** : Define relevant experiments in the Domain, as the basis for analogous Simulation Experiments and results analyses.

The relevant experiments take sections of prostate tissue, investigated under the microscope, and provide counts of the different cell types.

Domain Model (128) > Data Dictionary (132)

> **Data Dictionary (132)** : Define the modelling data used to build the simulation, and the experimental data that is produced by domain experiments and the corresponding Simulation Experiments.

We define the Data Dictionary, the specific data that is used to build the Domain Model:

- *Modelling data (parameter values)* : cell differentiation, division, and death rates; derived from the relevant sources in the Domain
- *Experimental data* : cell number ratios, #SC : #TA : #CB : #L, in a normal prostate; derived from the relevant sources in the Domain

The aim of the simulation here is to reproduce the cell number ratios in a normal prostate, before going on to model the propagation of abnormalities. Hence the experimental data comprises only Calibration (163) data; there is no need for further validation or unseen acceptance test data.

> The specific values of the ratios are omitted here. In the full development, they are recorded at this point, and links to the research data provided.

Domain Model (128) > Stochasticity Model (154)

> **Stochasticity Model (154)** : Model any required stochasticities explicitly.

Stochasticity arises from variations across tissue, and variations between samples. This is captured by the error bounds, variations and ranges acceptable on the cell number ratios, in the Data Dictionary.

Domain Model (128) > Document Assumptions (108)

Our prostate cell division and differentiation model abstracts away from low-level detail (cell signalling, intra-cell mechanisms, biochemistry, spatial aspects, physics, etc.) The Domain Model ignores concepts other than the identified and agreed cell types and transitions. In order to achieve a model that is computationally feasible, other concepts are simplified (for example, the information contained and transmitted in a cell's genome).

A.10 We can model the elastic constraint of the size of the prostate, an environmental "stop growing" signal, as a function of the total cell number, not the number or ratio of individual epithelial cell types; we can model this as a global constant that can be transmitted to all cells, as a "pressure surrogate".

   reason: In reality, the elastic constraint of the prostate is related to organ membrane physics: the complexity of this militates against its inclusion in the model (which is, in any case, aspatial); however, cancer is a consequence of out-of-control division, so a model that allows cell types to divide indefinitely would be unable to clearly distinguish cancerous from non-cancerous outcomes; furthermore, we need to be able to model cell death, and the elastic constraint is one of the recognised promoters of cell death, for example among the (mature) luminal cells.

justification: The modelling assumption and its consequences has been discussed and agreed by the Domain Scientist and Domain Modeller.

A.11 We can model division as a transition from one cell to two daughter cells (which then differentiate); this is an acceptable surrogate for natural division, which results in differentiated cells.

reason: For model simplicity: to allow separation of cell division and cell differentiation, without proliferating cases.

justification: This approach, suggested by the Domain Modeller, was discussed with and agreed by the Domain Scientist. Whilst we cannot systematically identify and count such daughter cells in biological assays, it is sometimes possible to identify cells that are in the process of division; it may be useful to be able to add delays or other factors to the daughter cells to account for the time taken to divide. Although there is no direct analogy in the biological system, this is close to how biologists mentally visualise the process. Furthermore, there is no biological data on the number of cells that are in the process of proliferation at any time.

consequence: This representation cannot readily capture the correlated asymmetric transition probabilities of the two daughter cells (with one being highly likely to remain a p1 cell, and the other being highly likely to differentiate to p2); such asymmetry could be devolved to some internal state, but we do not have a state machine model for these surrogates.

consequence: Because daughter cells are a modelling concept not a biological concept, daughter cells do not die.

A.12 We can model individual cell types with discrete substates.

reason: In the biological system, not all cells of a given type can divide or differentiate: there is a cell-level behaviour, as well as a system-level behaviour; the precise controls on division are not fully understood, but include intra-, inter- and extra-cellular factors. At the level of abstraction of our model, we need to be able to control when each cell divides or differentiates, as well as modelling overall rates of transition; for example, a stem cell is considered to be active or quiescent (Figure 3.7); it can die from either state, but it can only be considered for transition if it is in the active state.

justification: Discussed with and agreed between the Domain Scientist and Domain Modeller; the modellers rely on the domain scientists to identify which cell states (if any) are relevant. Separating the state diagram of cells from the Petri net model of the system maintains clarity, and allows the potential to influence division and differentiation at the individual cell level (that is, mutations that arise due to environmental damage to single cells).

A cell can die from any state, but the rate of cell death can only be estimated (and then calibrated). It is impossible to count "dead" cells in a real system, because there is no "dead cell", just chemicals emitted by "stressed" cells, which may be dying or damaged, and hard-to-detect residues of one or more dead cells; dead biological cells are not well-modelled as "things" with "types".

3.4 Development phase

CoSMoS Simulation Project (92) > Development Phase (96)

Development Phase (96)
Build the scientific instrument: produce a simulation platform to perform repeated simulation, based on the output of the Discovery Phase (95).

- revisit the Research Context (119)
- develop a Platform Model (149)
- develop a Simulation Platform (161)

The initial Discovery Phase (95) has focused firstly on the area of interest the Domain Scientist (101), capturing a specific view of their take on the Domain (123), and identifying hypotheses that could drive a Simulation Experiment (177). We now have an accepted Domain Model (128) that captures this view, and forms a bridge to the software engineering development.

The Development Phase now creates the Simulation Platform (161), by way of a Platform Model (149), using systematic, traceable software engineering processes.

This section outlines some of the key concepts, focusing on where the CoSMoS approach might be unexpected or unusual.

3.4.1 Revisit the research context

Development Phase (96) > Research Context (119)

The Research Context (119) should be revisited at the start of the Development Phase, to reflect on what has been learned during the Discovery Phase (95), and to check whether anything has changed due to that.

Resources. Due to resource limitations, we decided to implement only one differentiation stage (SC to TA), and only the Petri net layer, not the state layer. Further elaboration is left to a later increment. This means that the Simulation Purpose is reduced: the proportion only of SC to TA cells will be measured.

Apart from this, the Research Context is unchanged.

3.4.2 Develop a platform model

Development Phase (96) > Platform Model (149)

Platform Model (149)
From the Domain Model (128), develop a platform model suitable
to form the requirements specification for the Simulation Platform
(161).

- choose a Modelling Approach (111) and application architecture for the platform modelling
- develop the platform model from the Domain Model (128). In
  particular:

  - remove the Domain Behaviours (135)
  - develop the Basic Platform Model (151) from the Basic Do-
    main Model (130)
  - develop the Simulation Experiment Model (152) from the
    Domain Experiment Model (135)

- Document Assumptions (108) relevant to the platform model
- if necessary, Propagate Changes (168)

The Platform Model is an engineering development of the Domain
Model, and a step towards Simulation Platform construction. The
model is shaped by engineering design decisions, detailing the
implementation of the structures, behaviours and interactions
identified in the Domain Model in a way that naturally translates
to Simulation Platform technologies. This might dictate that some
concepts in the Domain Model are abstracted or simplified, to allow efficient implementation.

Platform Model (149) > Modelling Approach (111)

Modelling Approach (111) : Choose an appropriate modelling
approach and notation.

The Domain Model (128) is a Hybrid Model (212) that uses both Petri
nets and state diagrams. Seamless Development (214) would suggest

that we should continue to use these for the Platform Model. However, the team was unable to find a reliable Petri net execution model that could support millions of cell-processes and incorporated state diagram components.

Here we need to make formal the linkage between the Petri nets and state diagram. The descriptive linkage described in [63] is now formalised through the definition of a Domain Specific Language (228) (DSL), then using a model transformation approach from model-driven engineering (MDE) to develop the Simulation Platform (161) implementation. Our approach is non-standard, because we use two modelling languages that are already well-defined: our DSL customises existing languages, rather than starting from the beginning with domain concepts. For details of the underlying metamodelling and DSL definition, see [177].

This modelling approach produces a textual Platform Model that forms an intermediate step between the diagrammatic Domain Model (128) (amenable to review by the Domain Scientist (101)) and the Simulation Platform implementation.

Platform Model (149) > remove Domain Behaviours (135)

> High-level emergent Domain Behaviours identified in the Domain Model should be removed from the Platform Model, if the purpose of the research is to investigate the emergence of these behaviours from other model components (see Figure 1.3). This ensures that the Simulation Platform (161) does not *Program In the Answer* (170). It then requires there to be some means in the Results Model (174) to identify that the emergent behaviours have indeed emerged: here, the means to count cells of different types.

The Domain Behaviours are the desired cell number ratios as recorded in the Data Dictionary (132). These do not appear explicitly in the Platform Model, but are a consequence of transition probabilities.

The design decision is to use arbitrary probabilities to develop the Simulation Platform, and then to perform Calibration (163), using systematic variation of probability parameters until biologically-acceptable cell ratios are obtained.

Platform Model (149) > Basic Platform Model (151)

Basic Platform Model (151)
Build a detailed model of the basic platform concepts, components and processes.

- develop the Basic Platform Model (151) from the Basic Domain
  Model (130)
- as needed, develop the Stochasticity Model (154)
- as needed, develop the Space Model (155)
- as needed, develop the Time Model (158)

The move from domain model to platform model requires transforming the Petri net Basic Domain Model into a suitable UML Basic
Platform Model.

The change of modelling language requires us to put some effort into maintaining the Seamless Development approach. We have
to demonstrate the mapping between the Domain Model and the Platform Model, and show that this maintains the properties of the domain
model. We systematically map the Domain Model concepts using a
Domain Specific Language. The detailed mapping approach is documented in [63].

Stochasticity Model (154) : The design of the cell and transition prob-
    ability mechanisms is kept simple. We use a simple data structure
    (a list or array) that records each transition probability for that
    cell, and links to the transition probabilities associated with the
    place where the cell token is located.
Space Model (155) : The model is aspatial (assumption A.7).
Time Model (158) : The model of time is give by Petri net firing prob-
    abilities, as defined in the Stochasticity Model.

[detailed model omitted]

Platform Model (149) > Simulation Experiment Model (152)

> **Simulation Experiment Model (152)**
> Define relevant experiments in the simulation, analogous to domain experiments.
>
> - build a model to support running Simulation Experiment (177)s that are analogues of domain experiments
> - design a simulation experiment initialisation approach
> - design experiment instrumentation and logging

> The Platform Model also adds instrumentation and interfaces to allow observation (visualisation), user interaction, and recording of the eventual results, for using the simulation platform to run a Simulation Experiment.

Model: The domain experiments measure cell counts in sections of prostate tissue: the analogue is to measure simulated cell count ratios after the simulation has run.

Initialisation: The simulation starts with a population of cells in "normal" prostate ratios; this ratio should be maintained over time by the various transition rates.

Instrumentation: Cell number ratios to be extracted from the simulation. The simple solution here is to record the number of cells in each place at appropriate time points in the simulation.

!! **Future:** In later increments we will need consider cells in transition. One choice would be to complete transitions before the counts are made, to avoid "losing" the cells in transition, but this will need discussion with the Domain Scientist: it may be useful to count cells in transition as well. This decision will need making, documenting (Document Assumptions), and including in the Research Context of the relevant increment.

Platform Model (149) > Document Assumptions (108)

[omitted here]

Platform Model (149) > Propagate Changes (168)

Propagate Changes (168) : Ensure that changes in one part of the system propagate throughout, to ensure consistency.

The need for changes may be discovered when developing the Platform Model (149). This is the point where the focus of the modelling changes from domain to computational considerations. These considerations may expose areas that are over-ambitious, ill-defined, or otherwise inadequate, and that need to be changed. The change should be propagated through the Domain Model (128) as well as being incorporated in the Platform Model. This allows assumptions to be revisited, a check that the Simulation Purpose (121) is still achievable, and for the models to be made compatible.

There are no changes to propagate in this case.

3.4.3 Develop a simulation platform

Development Phase (96) > Simulation Platform (161)

Simulation Platform (161)
Develop the executable simulation platform that can be used to run the Simulation Experiment (177).

- choose an Implementation Approach (160) for the platform modelling, following the principle of Seamless Development (214) as much as possible
- coding
- testing
- perform Calibration (163)
- Document Assumptions (108) relevant to the simulation platform
- if necessary, Propagate Changes (168)

The Simulation Platform is an engineering development of the Platform Model (149). It encodes the Platform Model in software (and as appropriate, hardware) platforms on which Simulation Experiment (177)s can be performed.

The Simulation Platform defines a set of parameters (variables) that allow the encoded model to be manipulated. The parameters are derived from the Domain Model (128) and interpreted through the platform model, thus making the simulation platform accessible to domain scientists with knowledge of the domain model.

The software (and hardware) engineering development is non-trivial, both because of the inherent complexity of the simulated processes, and because of the need to be able to Argue Instrument Fit For Purpose (186). It is appropriate to put some thought into the choice of software engineering methods and techniques, as well as the target languages and program architecture.

Simulation Platform (161) > Implementation Approach (160)

Implementation Approach (160)
Choose an appropriate implementation approach and language.

- determine coding language, development environment, and approach
- determine which existing libraries and generic simulation frameworks might be used
- determine testing strategy

The approach used to develop the Simulation Platform needs to lead to an application that is traceable back to the Domain Model (128), is flexible enough to allow a range of related Simulation Experiment (177)s and hypotheses to be addressed, and for which it is possible to to Argue Instrument Fit For Purpose (186). The approach must also be feasible within the resources of the project as laid out in the Research Context (119).

Based on the team's existing expertise, and the principle of Seamless Development, we chose to use the Java object-oriented programming language. In engineering terms, this choice has the advantage of being able to use debugging and documentation support inherent in the Eclipse programming environment. In personnel terms, we had good experience with OO development.

We do not use any specific Agent Based Modelling libraries or frameworks. (In retrospect, a framework like Java Mason might have simplified development.)

[testing strategy omitted here]

!! Future: In later increments we aim to minimise the hand-coding activity, and maximise generation of code direct from Platform Model diagrams, or via intermediate model-driven engineering transformations.

Simulation Platform (161) > coding

The Simulation platform was coded in Java from the UML description of the Platform Model (149), using best software engineering practices.
   [code omitted here]

Simulation Platform (161) > testing

Testing was done according to best software engineering practices.
   [test cases and results omitted here]

Simulation Platform (161) > Calibration (163)

Calibration (163) : Tune the Simulation Platform so that simulation results match the calibration data provided in the Data Dictionary (132).

Domain (123) data values map to settings or parameter values in the simulator, through two translations, from Domain Model (128) to Platform Model (149), and then from Platform Model to Simulation Platform values. Some of these mappings are straightforward; others involve more complicated surrogating.

> Additionally, there are some values needed in the Simulation Platform that are not known from the Domain, and have to be estimated. Calibration tunes the Simulation Platform values. In order to Argue Instrument Fit For Purpose (186), the tuned values need to be mapped back to the Domain, and shown to have plausible values there, or shown not to be crucial through Sensitivity Analysis (175).

We know approximate rates of transition between states, and approximate time taken to divide and differentiate. The Domain Scientist can provide estimated distribution data for the probability of cell death. We can get some approximate data on the ratio of cells that are dividing relative to those that are in a particular "place".

We cannot get data that separates a "daughter cell" from its transitions, because the daughter cell is an modelling concept introduced to simplify and clarify the model of division and differentiation: see assumption A.11. We cannot get any information on the "sort" of dead cell (dead stem cell, dead luminal cell, etc.): see assumption A.12.

We calibrate the simulator's rates to achieve cell number ratios that are stable within the ranges for a normal prostate.

Simulation Platform (161) > Document Assumptions (108)

[omitted here]

Simulation Platform (161) > Propagate Changes (168)

No changes to propagate.

3.5 Exploration phase

CoSMoS Simulation Project (92) > Exploration Phase (97)

Exploration Phase (97)
Use the simulation platform resulting from the Development Phase
(96) to explore the scientific questions established during the Discovery Phase (95).

- initially, revisit the Research Context (119)
- develop an experimental Results Model (174)
- finally, revisit the Simulation Purpose (121)

The Exploration Phase pattern expansion is left incomplete below.
This allows us to illustrate how incompleteness can be handled
in CoSMoS Simulation Project documentation, and how it mani-
fests in the argumentation stage.

3.5.1 Revisit the research context

Exploration Phase (97) > Research Context (119)

The Research Context (119) for this phase is unchanged.

3.5.2 Develop a results model

Exploration Phase (97) > Results Model (174)

Results Model (174)
Build an explicit description of the use of, and observations from,
the Simulation Platform (161).

- perform Sensitivity Analysis (175)
- perform relevant Simulation Experiment (177)s
- build a Simulation Behaviours (179) model

The Results Model captures understanding of the Simulation Platform based on the output of simulation runs, and provides the basis for interpretation of what the simulation results show. Its relationship to the Simulation Platform is analogous to the relationship of the Domain Model (128) to the Domain (123).

The Results Model is constructed by experimentation and observation of Simulation Experiments, and might record observations, screen-shots, dynamic sequences, raw output data, result statistics, as well as qualitative or subjective observations. Sensitivity Analysis also provides input data to this process.

The contents of the Results Model are compared to the Domain Model to establish whether the Simulation Platform (161) provides a suitable representation of the real world Domain (123) being investigated. The results model might also provide details to develop new experiments, either on the Simulation Platform or in the real world Domain.

Results Model (174) > Sensitivity Analysis (175)

Sensitivity Analysis (175) : Determine how sensitively the simulation output values depend on the input and modelling parameter values.

[omitted]

Results Model (174) > Simulation Experiment (177)

Simulation Experiment (177)
Design, run, and analyse simulation experiments.

- design the experiment
- perform simulation runs and gather data
- analyse results, for input to the Simulation Behaviours (179) model
- Document Assumptions (108) relevant to the simulation experiment

---

Results Model (174) > Simulation Behaviours (179)

Simulation Behaviours (179)
Develop a model of the emergent properties of a Simulation Experiment (177), for comparison with the related emergent Domain Behaviours (135) of the Domain Model (128).

- build a *minimal* model, from consideration of the Research Context (119), the Simulation Experiment Model (152), the Domain Behaviours (135), and the Calibration (163) translation of the raw simulation data
- if needed, build an *augmented* model including micro-level observations, and argue the connection to the domain model data
- if needed, build a Visualisation Model (181)

The Simulation Behaviours model is the analogue of the Domain Behaviours (135) emergent properties model.

In this example, most of the work is done in Calibration, to ensure the cell ratios are correctly maintained by the transition probabilities. The minimal model is simply these ratios, plus the computational resource requirements. For a more realistic minimal model, see the larger case study in Part IV.

- minimal model : cell ratios
- augmented model : not required here
- Visualisation Model (181) : not required here

3.5.3 Revisiting the simulation purpose

Exploration Phase (97) > **Simulation Purpose** (121)

The overall aim of the project is to: develop a model and simulation of prostate cell differentiation and division, based on prostate cell populations from laboratory research data.

This has been achieved: we have a Domain Model agreed by the Domain Scientist, which we have developed into a Platform Model and a fit-for-purpose Simulation Platform. By running Simulation Experiments, we have built an appropriate Simulation Behaviours model.

We have a fit-for-purpose simulator that is documented. The next increment can start from this point, rather than from scratch. We can modify and extend these models and simulator to answer more important questions.

3.6 Argue instrument fit for purpose

CoSMoS Simulation Project (92) > **Argue Instrument Fit For Purpose** (186)

Argue Instrument Fit For Purpose (186)
Provide an argument that the CoSMoS Simulation Project (92) simulation is fit for purpose.

- establish the fitness-for-purpose *claim*, from the intended purpose of the simulation, as recorded in the Research Context (119)
- establish the required *rigour* of the argument, as recorded in the Simulation Purpose (121)
- agree a *strategy* for substantiating the fitness-for-purpose claim
- use a Structured Argument (189) to substantiate the fitness-for-purpose claim

The final part of the development (presentationally, if not pro-
cedurally) is the argumentation.

To establish fitness for purpose, it is necessary to create an argument that an appropriate instrument has been designed, with
respect to its purpose. There should be an argument that justifies
the trust in the Simulation Platform, its use to run Simulation Experiments, and the results it delivers. Fitness for purpose can be
presented as an argument that the simulation meets its scientific
and engineering objectives. The argument is open to scrutiny by
whoever needs convincing.

Where the Simulation Purpose relates to support for laboratory research, the argument should also consider the adequacy of
results of simulation, and their comparison to results of similar
laboratory experiments.

Argue Instrument Fit For Purpose (186) > claim

The Research Context's Simulation Purpose for this increment is to develop a model and simulation of prostate cell differentiation and division, where

1. the Domain Model is agreed suitable by the Domain Scientist
2. the simulation replicates observed cell population dynamics, represented as changing proportions of cells in a "normal" prostate
3. the simulation can be run at a sufficient large scale to investigate
   low probability events (mutations of single cells)

The claim is that the developed simulation is fit for this purpose.

Argue Instrument Fit For Purpose (186) > rigour

The Simulation Purpose states: "As a feasibility study, the specific simulation results are non-critical, and so lightweight arguments of fitness for purpose are all that is needed for these. As the purpose of
this phase is to decide whether to proceed with a full simulation, argumentation related to scaling properties and resource requirements
is more critical."

Argue Instrument Fit For Purpose (186) > strategy

The strategy for establishing this claim is to Use Generic Argument, by suitably instantiating the example generic template given in Create Generic Argument, and arguing the relevant subclaims using a Structured Argument.

Argue Instrument Fit For Purpose (186) > **Structured Argument** (189)

Structured Argument (189) : Structure and develop the required arguments in a systematic manner.

Here we show the top level of the argument, in tree form, adapted from [176].

This tree form is used to show a summary of the structure of the argument, and act as an index into the detailed body of the argument, cross referenced by the various identifiers (below).

Claims annotated with a black diamond ◆ are presented separately elsewhere (below); claims annotated with a white diamond ◇ are not further developed; claims annotated with a ⍁ have accompanying separately textual commentary (below).

[**claim** 1] the prostate simulation is fit for purpose
    [**context** 1.1] Simulation Purpose documents role and criticality
    [**context** 1.1] CoSMoS Simulation Project documents the models and simulator
    [**strategy** 1.2] argue over: (i) the scientific domain; (ii) the implementation; (iii) the experiments; (iv) the interpretation of the results
        [**justification** 1.2.1] the team agree that this strategy is sufficient for the stated purpose
        [**claim** 1.2.2] scientific domain: the Domain Model adequately captures the Domain for the simulation purpose ⍁
            [**strategy** 1.2.2.1] argue over the model content and assumptions
                [**context** 1.2.2.1.1] the documented assumptions
                [**justification** 1.2.2.1.2] sign-off from the relevant stakeholders
                [**claim** 1.2.2.1.3–N] ◇

[**claim** 1.2.3] implementation: the implementation adequately captures the Domain Model for the simulation purpose

> [**strategy** 1.2.3.1] argue over (i) the derivation of the Platform Model from the Domain Model, and (ii) software engineering, testing, and calibration of the Simulation Platform
>
> > [**claim** 1.2.3.1.1] derivation: the Platform Model is adequately derived from the Domain Model ◆
> >
> > [**claim** 1.2.3.1.2] software engineering: the Simulation Platform is adequately engineered ◆
>
> [**claim** 1.2.4] experiments: the Simulation Experiment is adequately performed ◇
>
> > [**strategy** 1.2.4.1] argue over use within calibration, random seeds, results analysis, and comparison with domain results
>
> [**claim** 1.2.5] interpretation of results: the Simulation Behaviours are adequately related to the Domain Behaviours ◇

Claim 1.2.2: the Domain Model adequately captures the Domain for the Simulation Purpose

> An argument that the Domain Model is fit for purpose is based on the explicit statement of the Simulation Purpose (121).

The Simulation Purpose is to explore the dynamics of cell division and differentiation in normal prostate and cancer neogenesis, in such a way that mutation and heritability effects could be demonstrated.

This first increment is an initial feasibility study, to demonstrate whether the simulation approach is computationally feasible.

Many cancer domain scientists would argue that intentionally ignoring major aspects of neogenesis (such as angiogenesis, hormonal signalling, and cell-cell contacts) will render the model useless. This shows a lack of understanding of the power of the modelling approach, which is to allow focussed questioning of a part of the system, rather than attempting to put in all of the detail. By over-complicating the simulation, no single part of it would be amenable to analysis, and the effort would be wasted. Implicit in the creating of intentionally incomplete models is that there is a lot more going on than is present in the simulation. This will always be the case, but is not necessarily detrimental, as long as the model is not used outside its design limitations.

During domain modelling, we have presented a range of comments on the appropriateness of modelling notations, scope and scale of modelling, and the content of the models. This commentary contributes to the Research Context, and also provides the basis for arguing that the domain model is fit for purpose.

Claim 1.2.3.1.1: the Platform Model is adequately derived from the Domain Model

As in conventional requirements analysis, the most obvious, and most difficult to bridge, semantic gap in software engineering is between the Domain and the abstract model or specification (Platform Model). By using well-defined software engineering notations, and principles of model-driven engineering such as Seamless Development (214), we can significantly reduce the risk of changing the intention of the simulator during development: since the Domain Model (128) is argued fit-for-purpose, then the best practice software engineering processes employed should ensure that the implementation is also fit for purpose.

Removals: The Domain Behaviours of the cell number ratios are not included in the Platform Model.

Changes: There was no need to make changes (for example, surrogating) in moving between models.

Translation: The translation from Petri net to UML is correct. A Domain Specific Language is used to automate the Petri net-to-UML translation. The approach has been published and peer-reviewed [63].

Claim 1.2.3.1.2: the Simulation Platform is adequately engineered

We use the **!! Unfit** tag to mark aspects of the argument that indicate a weakness or omission discovered in the development, indicating the instrument may be Unfit (199) for purpose.

Engineering: The translation from UML to Java has used best software engineering practices.

Testing: Adequate testing has been performed.

!! **Unfit:** *The resulting ratios are somewhat sensitive to parameter values (according to a third party reviewer).* Sensitivity Analysis *needs to be performed prior to further simulation effort.*

Calibration: Calibration has adjusted the various rates to get the pre-determined cell number target ratio.

!! **Unfit:** *A further argument is needed that the pre-determined target ratio used in calibration is biologically realistic. It is likely that simulation experiments will need to be designed to explore the stability of different cell number ratios within the range that the Domain Scientist considers to be indicative of a normal prostate.*

> The argument here has exposed two areas where more development work may be needed. This needs to be taken back to the project team, who could decide that the argument as it stands is sufficient, or that the extra work is needed before results can be agreed, or that the extra work can wait until a later increment.

3.7 Real world simulation

The example in this chapter demonstrates the basic CoSMoS process. Part II gives detail of the different patterns making up this basic process, and Part III gives further patterns suitable for more specialised, or larger, projects. Part IV provides a much larger, fully worked through, CoSMoS case study. On the other hand, CoSMoS concepts can also be used in much smaller projects to help organise the components, for example, [1].

Part II
The Core CoSMoS Pattern Language

A CoSMoS Simulation Project tends to involve three (not necessarily sequential) phases, each with different motivations, activities, and focus.

The following chapters provide patterns for the various phases, and form the "core" CoSMoS approach. Start your project at the CoSMoS Simulation Project pattern, and follow the guidance, using the other referenced patterns as your specific context demands. The development and use of a simulation as a scientific instrument will probably include most of these core patterns.

Ch.4 The top level CoSMoS Simulation Project (92) pattern: phases, roles, and generic patterns

Ch.5 Discovery Phase (95) patterns: determining the Research Context (119), scoping, exploration, prototyping, building the Domain Model (128)

Ch.6 Development Phase (96) patterns: building the Platform Model (149), software engineering the Simulation Platform (161)

Ch.7 Exploration Phase (97) patterns: using the Simulation Platform (161) as a scientific instrument to perform Simulation Experiment (177)s

Ch.8 Argumentation patterns: how to Argue Instrument Fit For Purpose (186) using a Structured Argument (189)

Each chapter is split into sections for different aspects of its topic: the first section is a catalogue of included patterns, and subsequent sections present each full pattern using the pattern language and templates defined in Chapter 1.

Chapter 4
The CoSMoS simulation pattern

Abstract — In which we describe the structure of the overall CoSMoS simulation phase patterns.

4.1 Catalogue of top level patterns

Phase patterns	
CoSMoS Simulation Project (92)	Develop a fit for purpose simulation of a complex scientific or engineering system.
Discovery Phase (95)	Decide what scientific instrument to build: establish the scientific basis of the project, identify the domain of interest, model the domain, and shed light on scientific questions.
Development Phase (96)	Build the scientific instrument: produce a simulation platform to perform repeated simulation, based on the output of the Discovery Phase (95).
Exploration Phase (97)	Use the simulation platform resulting from Development to explore the scientific questions established during the Discovery Phase (95).

S. Stepney, F.A.C. Polack, *Engineering Simulations as Scientific Instruments: A Pattern Language*, https://doi.org/10.1007/978-3-030-01938-9_4

Role patterns

Roles (99)	Assign team members to key roles in the simulation project.
Domain Scientist (101)	Identify the lead provider of Domain (123) knowledge.
Domain Modeller (103)	Identify the lead developer of the Domain Model (128) and Results Model (174).
Simulation Engineer (105)	Identify the lead developer of the Platform Model (149) and Simulation Platform (161).
Argument Modeller (106)	Identify the lead developer of the fitness-for-purpose argument.

Generic patterns

Document Assumptions (108)	Ensure assumptions are explicit and justified, and their consequences are understood.
Modelling Approach (111)	Choose an appropriate modelling approach and notation.

Antipatterns

Box Ticking (??)	You are blindly following the process as a bureaucratic box-ticking exercise.
Cutting Corners (115)	You follow the CoSMoS approach initially, but you do not revisit, update, or iterate.
Divorced Argumentation (115)	You do the argumentation as a separate exercise.

4.2 Phase patterns

CoSMoS Simulation Project

Intent

Develop a (single-increment) fit for purpose simulation of a complex scientific domain or engineering system.

Summary

- carry out the Discovery Phase (95)
- carry out the Development Phase (96)
- carry out the Exploration Phase (97)
- Argue Instrument Fit For Purpose (186)

Context

This is the top level pattern, for doing (one iteration of) a CoSMoS project.

Discussion

CoSMoS provides a way to bring software engineering best practice into the development of simulations. In common with conventional software engineering lifecycles (e.g. Royce's original Waterfall, Boehm's Spiral, Sargent's simulation development), CoSMoS separates concerns in modelling.

When concentrating on the Domain (123) (the subject of simulation), create models that abstract from the domain of study, and do not include implementation details.

On moving to the Platform Model (149), map from Domain concepts to software engineering concepts (e.g. in an object-oriented platform context, map from cells to objects, from types of cells to classes). Here, add concepts that are irrelevant to modelling the Domain but essential to the implementation, such as ways to collect, save, and perhaps visualise data.

On proceeding to creation and calibration of the Simulation Platform (161), successively convert more of the Platform Model to a concrete implementation model, which contains successively more of the implementation concepts (foe example, in an object-oriented context, add setter and getter methods on classes; in a concurrent context, add synchronisation mechanisms; in a process-oriented context, determine mobile and static processes, channels and protocols).

The guiding principles are: (a) to use only the terms and concepts that are appropriate to the audience at each stage; (b) to explicitly document the concepts that persist from the domain into the implementation, so that it is always clear how a domain concept is implemented (Seamless Development (214), with traceability).

The three phases provide different emphases: on understanding the domain, on building the platform, on using the platform. These phases are not intended to be rigidly sequential: earlier phases may be revisited as more is uncovered in later phases; models may be refactored and refined; fit-for-purpose arguments help glue the phases together. Indeed, sometimes Discovery Phase modelling is all that is needed or possible. For example, you may discover something that you take forward via a different route, or discover that you do not have the resources or knowledge needed to progress.

Related patterns

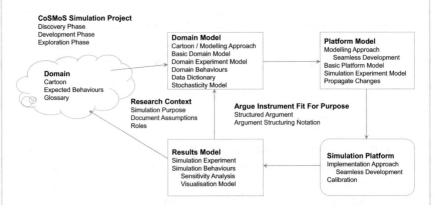

Fig. 4.1 The core patterns involved in a CoSMoS Simulation Project

The core patterns involved across the three phases are shown in Figure 4.1.

As the scale or complexity of your simulation project grows, you may also need to consider one or all of Multi-increment Simulation (240), Multi-domain Simulation (242), and Multi-scale Simulation (244).

If you have an Engineered Domain (238), rather than a natural domain, the CoSMoS approach can still be used, but the results need to be interpreted differently.

You can tailor the overall process for your particular project; some examples are discussed in Partial Process (230) and Post Hoc (232).

Discovery Phase

Intent

Decide what scientific instrument to build. Establish the scientific basis of the project: identify the domain of interest, model the domain, and shed light on scientific questions.

Summary

- identify the Research Context (119)
- define the Domain (123)
- construct the Domain Model (128)

Context

The first phase of the top level CoSMoS Simulation Project (92).

Discussion

The driving force behind the Discovery Phase is the need to understand and scope the research that is going to be conducted on the engineered Simulation Platform (161). The goals, in relation to scientific research simulation, are:

- to identify the scientific basis for a CoSMoS project, establishing the domain of interest
- to understand the domain of interest and capture a model of this understanding
- to establish a set of questions to ask of the domain model via simulation

Given these goals, the Discovery Phase results in the identification of the Domain (123) and the creation and modification of the Domain Model (128) and Research Context (119).

Related patterns

Through interaction with the Domain Scientist (101), a Domain Modeller (103) gains understanding of part of the Domain (123). In later phases, this understanding can feeds into Simulation Experiment (177), which should then be discussed and agreed with the Domain Scientist (101), so that the Research Context (119) is mutually understood.

Development Phase

Intent

Build the scientific instrument: produce a simulation platform to perform repeated simulation, based on the output of the Discovery Phase (95).

Summary

- revisit the Research Context (119)
- develop a Platform Model (149)
- develop a Simulation Platform (161)

Context

The second phase of the top level CoSMoS Simulation Project (92).

If the phases are being performed in the standard order, there will be a Domain Model (128), expressed in a well-defined language, that has been accepted by the Domain Scientist (101) as an acceptable representation of the part of the Domain (123) to be captured in the Simulation Platform (161).

Discussion

The purpose of the development phase is to engineer a Simulation Platform (161) upon which to carry out the scientific research identified in the Discovery Phase (95). The Development Phase encompasses two aims:

- to transform the Domain Model (128) into a Platform Model (149) that describes how to undertake the simulation-based research identified in the Discovery Phase (95)
- to develop the Platform Model (149) into a Simulation Platform (161) of appropriate quality, flexibility and reliability

Depending on the criticality and impact of the intended use of the simulation results, the Development Phase may be agile, lightweight, or a rigorous engineering exercise. In most cases the intended use of simulation results is to explore a scientific question as part of a wider set of scientific research goals and research activities.

The pattern includes the option to revisit the Research Context (119), as more discoveries are made during development.

Related patterns

At the end of the development phase, the Simulation Platform (161) exists, and should be in a form suited to the research objectives identified in the Discovery Phase (95). The Simulation Engineer (105) should be able to give suitable undertakings on the quality of the engineering – appropriate unit testing and other quality-enhancing activities should have been undertaken. This separates the engineering aspects of validation from the scientific aspects of validation, which relate to whether the simulation has anything meaningful to say about the Domain (123).

Exploration Phase

Intent

Use the Simulation Platform resulting from the Development Phase to explore the scientific questions established during the Discovery Phase.

Summary

- initially, revisit the Research Context (119)
- develop an experimental Results Model (174)
- finally, revisit the Simulation Purpose (121)

Context

The third phase of the top level CoSMoS Simulation Project (92).

If the phases are being performed in the standard order, there will be a tested Simulation Platform (161), in need of Calibration (163), and suitable for use to run a Simulation Experiment (177).

Discussion

The exploration phase uses the Simulation Platform to address the scientific research identified during Discovery Phase. This is where the relevance of the simulation results has to be addressed. To do this, and to complete the research objectives that led to simulation, a number of specific simulations may be derived from the platform (with or without the need to change the Platform Model or Research Context).

The Exploration Phase applies both to the generic Simulation Platform, and to the specific Simulation Experiment. The phase can be summarised as:

- performing Simulation Experiment (177) on the Simulation Platform (161) to generate results
- building a Results Model (174) including a Simulation Behaviours (179) model, by assessing outputs and analysing results: evaluation, scientific validation, etc.

There are two sorts of output from the exploration phase: discoveries about the simulation and discoveries from the simulation. The former relate to the adequacy of the results; the latter contribute to the Results Model.

Discoveries about the simulation relate to scientific validation activities of the Exploration Phase and determine whether the simulation produces qualitatively similar results to the Domain Behaviours. Even if the simulation is a good engineering product (the endpoint of the Development Phase), mismatches have many possible causes:

- The Development Phase produces a Simulation Platform that does not adequately capture the behaviours or interactions of the agents and/or the environment. This points to issues relating to realisation of the Domain Model: the simulation may be sound, but the design decisions or assumptions are at fault.
- The Domain Model contains misunderstandings or inappropriate abstractions from the Domain. This focuses on issues relating to the assumptions and decisions taken in arriving at a mutually-understood and sufficient Domain Model. The problems could relate to the understanding of active components of the system, to understanding of the environment, and/or to understanding of interactions. Furthermore, there are scale issues: failure to produce emergent behaviours may result from the wrong (absolute or relative) numbers or types of agents, the wrong relative sizes of components, and many other possible failures or omissions.
- In the Domain, there may be fundamental misunderstandings. In this case, the Simulation Platform may be doing exactly what it is intended and designed to do, but the science identified by the Domain Scientist is incorrect.

The interpretation problem is compounded by the likelihood of more than one sort and instance of these causes. Issues uncovered during scientific validation often prompt further increments of the CoS-MoS process. The process is similar to that of laboratory science: a continual querying of the evidence and the basis of the evidence, until reasonable certainty is achieved.

Discoveries from the simulation are the results. These may be data or visualisations that can be used to complement scientific research, or they may be qualitative understandings of what processes might or might not be involved in certain complex behaviours. The discoveries must always be considered in the context of the Domain Model, the engineering (development phase products) and the Domain. In particular, they should be analysed and evaluated in the context of the Simulation Purpose.

4.3 Role patterns

Roles

Intent

Assign team members to key roles in the simulation project.

Summary

- identify the Domain Scientist (101)
- identify the Domain Modeller (103)
- identify the Simulation Engineer (105)
- identify the Argument Modeller (106)
- identify other *optional roles*
- identify necessary *collaborations* between roles

Context

A component of the Research Context (119) pattern.

Discussion

The team members need to ask certain questions and perform certain tasks in order to fulfil the intent of the various patterns. To do this, they take on specific roles, which act as a prompt for those questions and tasks. It is important to be able to split the science (that the instrument will investigate) and the engineering (the design, building and deployment of the instrument).

The key roles of Domain Scientist, Domain Modeller, Simulation Engineer, and Argument Modeller are core to every use of the CoSMoS approach. Roles indicate *responsibilities*, not *people*. We are not suggesting that there is one person per role on the team: one person may undertake many roles during the project; one role may be undertaken by many people; the person taking on the lead role may change with context or over time.

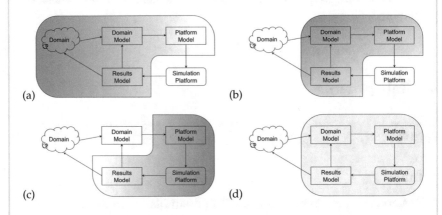

Fig. 4.2 Where the different key roles' effort is focussed during the project: (a) Domain Scientist; (b) Domain Modeller; (c) Simulation Engineer; (d) Argument Modeller. See specific role patterns for more detail

Other (sub)roles are useful to designate. The purpose of identifying the roles is to help clarify what needs to be done, when. Such roles might include:

Scribe: Knowledge elicitation meetings can be very productive, but only if the knowledge (facts, opinions, data, sources, assumptions, and so on) is captured. A designated Scribe ensures this occurs. Depending on the criticality and size of the project, the Scribe's task might range from capturing whiteboard contents to recording and transcribing interviews.

Experimenter: The Simulation Platform user performing the Simulation Experiment in order to explore the Domain (123). For small projects, this will typically be the same person as the holder of the Domain

Scientist role. For larger projects, particularly if the simulation is being used as the basis of a larger experimental programme, the experimenter may be a scientist more remote from the original Domain Scientist.

Stakeholder representative: There may be external stakeholders involved in the project, such as policy-makers, safety evaluators, simulation maintainers, and more.

A CoSMoS Simulation Project involves team members with diverse backgrounds, areas of competence, and work practices working closely together. It is crucial that the team members should respect each others' different competencies and priorities. They should realise that different disciplines have different constraints (for example, the different time-scales of wet lab and *in silico* experimental runs).

Team members should develop trust in the input from other members with different expertises. Trust should foster an atmosphere where naive questions can be asked with confidence, and taken seriously.

In addition to mutual trust between team members, trust needs to be built in the Simulation Platform, and in the Simulation Experiment results. The CoSMoS approach helps develop that trust, through the use of disciplined development, and the fitness-for-purpose argumentation.

Related patterns

Beware the role that leads to *Amateur Science* (137). Despite trust, avoid the traps of *Uncritical Domain Scientist* (143) and *Uncritical Domain Modeller* (143).

Domain Scientist

Intent

Identify the lead provider of Domain (123) knowledge.

Summary

- determine the required expertise for the role, within the Research Context (119)
- identify the team members responsible for providing domain knowledge
- agree the lead member, the problem "owner"

Context

A component of the Roles (99) pattern.

Discussion

The Domain Scientist provides all the information (on process, structures, inputs, expected results, and so on) necessary to create a fit-for-purpose simulator. The Domain Scientist is usually the owner of the Research Context (119) and may be the one who designs the Simulation Experiments to be run on the resulting Simulation Platform.

Determining the required expertise for the role leads to discussion of the limitations of Domain understanding and the causes of limited understanding, exposing further assumptions.

Literature is dense, incomplete, inconsistent, potentially irrelevant, and error-prone. It is not a sufficient resource on its own for building a simulation (see *Amateur Science* (137)). The Domain Scientist provides the necessary interpretation of the literature, and can indicate what is pertinent, which route to take in the presence of conflicting results, and where data is unavailable and what assumptions can be made in this case.

On anything but the smallest project, there will be multiple sources of input, and multiple (sub)Domain Scientists. But it is essential to identify a single individual "problem owner", to arbitrate on conflicting information.

This role should still be called the Domain *Scientist*, even in an engineering domain. This emphasises that the role requires a scientific stance, of hypotheses, experiments, and such like, rather than a design stance.

Fig. 4.3 The Domain Scientist role effort is focussed on the Domain, on providing information for the Domain Model, on agreeing changes made for the Platform Model, and on providing information to ensure the Results Model is comparable with real world experiments

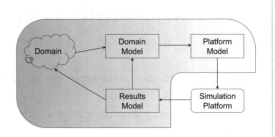

Related patterns

A Multi-domain Simulation (242) in particular will need several Domain Scientists providing input on potentially quite divergent domains, to provide the necessary material.

The Domain Modeller (103) must collaborate with the Domain Scientist, else fall into the trap of *Amateur Science* (137). Avoid the traps of *Uncritical Domain Scientist* (143) and *Uncritical Domain Modeller* (143).

Domain Modeller

Intent

Identify the lead developer of the Domain Model (128) and Results Model (174).

Summary

- determine the required expertise for the role, within the Research Context (119)
- identify the team members responsible for performing domain modelling
- agree the lead member, the domain model "owner"

Context

A component of the Roles (99) pattern.

Discussion

The Domain Modeller converts the Domain (123) information produced by or via the Domain Scientist (101) in to an appropriate Domain Model (128) and associated Results Model (174). The key responsibilities of the Domain Modeller include:

- identifying an appropriate Domain Modelling Approach (111) that can be taken forward to implementation, in consultation with the Simulation Engineer
- identifying appropriate levels of abstraction, scope and scale
- taking decisions about the design and representation of pertinent features of the Domain

- building the Domain Model from Domain information, in consultation with the Domain Scientist
- being an interface between the Domain Scientist and the Simulation Engineer, to ensure that the implementation remains consistent with the agreed Domain Model

Determining the required expertise for the role motivates discussion of appropriate abstraction and modelling approaches.

Separation of the role of Domain Modeller focuses attention on the obligations and responsibilities that pertain to the capturing domain knowledge both in a form understandable to the Domain Scientist, and in a form suitable for taking forward into simulation by the Simulation Engineer. The Domain Modeller needs to avoid simplifying the Domain for future implementation reasons, and must consider the validity of each design decision that is taken. It is important for the Domain Modeller to understand that every step of the development has the potential to corrupt or invalidate the simulation as a scientific or engineering instrument.

Working with the Simulation Engineer (105) and Domain Scientist, the Domain Modeller is also responsible for devising the Calibration (163) of the Simulation Platform (161), by the selection of appropriate data and scenarios with which to tune the simulation platform, to demonstrate a reliable base-line simulation.

Fig. 4.4 The Domain Modeller role effort is focussed on building the Domain Model, on the changes made for the Platform Model, and on building the Results Model

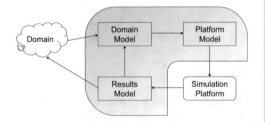

Related patterns

The Domain Modeller should collaborate with the Argument Modeller (106) to initiate the process to Argue Instrument Fit For Purpose (186), to ensure that the Domain Model is accepted as fit for purpose by the Domain Modeller and the Domain Scientist.

Simulation Engineer

Intent

Identify the lead developer of the Platform Model (149) and Simulation Platform (161).

Summary

- determine the required expertise for the role, within the Research Context (119)
- identify the team members responsible for performing platform modelling and simulator implementation
- agree the lead member, the platform "owner"

Context

A component of the Roles (99) pattern.

Discussion

The Simulation Engineer is responsible for building the Platform Model and implementing the Simulation Platform, including the verification of the code using suitable software engineering approaches. The Simulation Engineer converts the Domain Model (128) produced by the Domain Modeller (103) into an appropriate executable Simulation Platform. The key responsibilities of the Simulation Engineerinclude:

- identifying an appropriate platform Modelling Approach (111) that can provide a bridge between the Domain Model and the Simulation Platform implementation
- making, justifying and recording any simplifications needed to accommodate computational resource constraints, in consultation with the Domain Modeller and Domain Scientist
- taking design decisions about the implementation of all relevant features of the Platform Model and Visualisation Model (181)
- identifying the appropriate implementation media (languages, libraries, coding environments, hardware platform, etc.)
- identifying appropriate software engineering verification, for instance, writing test plans and test suites
- performing Calibration of the simulator, by running experiments using selected data and scenarios, and modifying the Simulation Plat-

form, to demonstrate a reliable base-line simulation, in consultation with the Domain Modeller

Determining the required expertise for the role motivates discussion of appropriate implementation and testing approaches.

Separation of the *role* of Simulation Engineer focuses attention on the obligations and responsibilities that pertain to the implementation of the Simulation Platform. The Simulation Engineer needs to create a suitably faithful implementation of the Domain Model, in as seamless a manner as possible, and needs to understand and record the mappings from the Domain Model to the Simulation Platform, reviewing the validity of each implementation decision that is taken. It is important for the Simulation Engineer to understand that every step of the implementation has the potential to corrupt or invalidate the simulation as a scientific or engineering instrument.

Fig. 4.5 The Simulation Engineer role effort is focussed on the design of the Platform Model, and on implementing the calibrated Simulation Platform from it

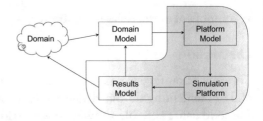

Related patterns

The Simulation Engineer should collaborate with the Argument Modeller (106) to extend application of the pattern Argue Instrument Fit For Purpose (186), to ensure that the Simulation Platform is accepted as fit for purpose by the Domain Modeller and the Domain Scientist.

Argument Modeller

Intent

Identify the lead developer of the fitness-for-purpose argument.

Summary

- determine the required expertise for the role, within the Research Context (119)
- identify the team members responsible for developing and documenting the fitness-for-purpose arguments of the various models
- agree the lead member, the argument "owner"

Context

A component of the Roles (99) pattern.

Discussion

The Argument Modeller is responsible for facilitating and capturing consensus among the team members on what is to be modelled, how it is modelled, and how it is implemented. Typically, all team members are involved at some point, but it is useful to have a lead person identified, who is responsible for capturing the fitness-for-purpose argument and presenting it to the rest of the team for review, revision and extension as appropriate.

Determining the required expertise for the role motivates discussion of appropriate argumentation approaches and detail.

The key responsibilities of the Argument Modeller include:

- to Document Assumptions (108) as they arise, and get agreement on their reasons, justifications, and consequences
- to get agreement on how the fitness-for-purpose argument will be represented (for example, the prostate study uses Structured Argument (189), represented using Argument Structuring Notation (190))
- to determine the criticality of the simulation, and thus the extent to which the Claim (192) of fitness for purpose needs to be substantiated: in a simulation deemed highly critical, they would be necessary to also identify the review process to be applied to the claim of fitness for purpose
- to determine a strategy for arguing fitness-for-purpose (for example, in the prostate study we decided to Use Generic Argument (202), which presents a high-level Strategy (194) that breaks the problem into three elements: capturing the domain, software engineering quality, and results analysis)

- to use the strategy to challenge the Domain Scientist, Domain Modeller and Simulation Engineer, extracting their understanding of why any Claim (192) is acceptable

Fig. 4.6 The Argument Modeller role effort is focussed on demonstrating that the four key artefacts of the project have been designed and implemented in a manner that is fit for purpose

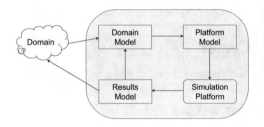

Related patterns

The Argument Modeller should collaborate with the Domain Scientist (101), Domain Modeller (103), and Simulation Engineer (105), to Argue Instrument Fit For Purpose (186), to ensure that the Simulation Platform (161) is accepted as fit for purpose by the Domain Scientist (101).

4.4 Generic patterns

This section documents some generic patterns that are relevant across all phases.

Document Assumptions

Intent

Ensure assumptions are explicit and justified, and their consequences are understood.

Summary

- identify that an assumption has been made, and record it in an appropriate way
- for each assumption, determine its nature and criticality
- for each assumption, document the reason it has been made

- for each reason, document its justification, or flag it as "unjustified" or "unjustifiable"
- for each assumption, document its connotations and consequences
- for each critical assumption, determine the connotations for the scope and fitness-for-purpose of the simulation
- for each critical assumption, achieve consensus on the appropriateness of the assumption, and reflect this in fitness for purpose arguments
- revisit the Research Context (119) in light of the assumption, as appropriate

Context

Any situation in which abstraction or simplification arises; representations are chosen; modelling, implementation or experimentation is undertaken, including: Research Context (119), Domain (123), Domain Model (128), Platform Model (149), Simulation Platform (161), Simulation Experiment (177). It is a responsibility of holders of all the relevant Roles (99) to make explicit the assumptions they may be making; it is the responsibility of the Argument Modeller (106) to ensure these assumptions are fully documented and agreed upon.

Discussion

A simulation that aims to support scientific research makes many simplifications and abstractions from the Domain, and may be based on uncertain scientific data. Further design decisions are needed to represent a continuous complex domain in a discrete digital computation. The assumptions that are made need to be explicit and documented. Consideration needs to be given to the implications of assumptions.

We can never hope to capture all the assumptions, or all the consequences of an assumption, but the more we capture, the better our understanding of the limitations of the simulation and the results that it produces. Assumptions may be identified as they are made, but are often identified retrospectively, and are often uncovered when people Ask [Silly] Questions (217), and during argumentation.

The identified assumptions are important in determining the fitness-for-purpose of the simulation, understanding the scope and capabilities of the simulation, and in interpreting results. So these assumptions need to be stated explicitly, the reasons given, and justified. Documented assumptions allow different team members with different assumptions to disagree, with the points of disagreement documented.

Reasons could include such things as "standard practice in the research domain", "to enable comparison with other results", "a consequence of a modelling decision or other assumption", "for ease of implementation", "for computational tractability", "this is the only data we have", "this is only the first increment of development, and will be revisited later", and so on.

The **justification** is relative to the Research Context (119) and Simulation Purpose (121) and must be checked if these change. Justifications could include such things as "this number of agents is representative", "the simulation result is insensitive to the precise assumption made", "appropriate given the research context". An assumption may be unjustified, and may even be unjustifiable. The justification may require a further argument to establish it, and may include reference to the literature that back up the justification.

Assumptions have **consequences**. Is the assumption critical to the development of a simulator? What are the consequences for the scope and fitness-for-purpose of the simulation? Does the assumption affect the interpretation of results? What are the importances of the consequences? The consequences may have an impact on the Research Context (119) and Simulation Purpose (121), for example, demonstrating that the stated purpose cannot be achieved under the assumption. For this reason, it is important to consider the potential consequences of each assumption.

Even for **low criticality** results requiring only lightweight argumentation, it is good practice to Document Assumptions, and to provide at least the reasons; justifications and consequences should be included where non-obvious. The Domain Scientist or Domain Modeller may make a simplifying assumption that is later found to be unnecessary or counterproductive. The Simulation Engineer may make an assumption to ease implementation issues that is later found to contradict or limit the use of the simulator for its intended Simulation Purpose. Assumptions justified at one time may become invalid later, in the light of new Domain knowledge, or new understanding of the Research Context, or if the Simulation Purpose changes.

As everywhere in the CoSMoS approach, documenting assumptions does not happen in one discrete, well-defined place. These patterns express the desired outcome, not the process by which it was obtained. Documenting assumptions and examining their consequences should ideally happen as soon as they are identified. Assumptions may nevertheless be uncovered late in the project, and may lead to backtracking, and possibly redefining of the scope. This can be an important and valuable part of the scientific endeavour. Conversation between the team members, and iteration through the process, should help reaching a usable compromise, and a revised Research Context. In this case, the documented justification might end up as "fits the research context": the assumption is justified as fitting the research context that has been modified in light of the uncovered assumption. In such a case, it is important to add a further assumption to the modified Research Context giving the full justification for the modification, so that the reasoning does not become lost.

Publication of results based on simulation should include documented assumptions (for example, in supplementary material). Readers can then understand the limitations of the results and their interpretation, such as what can and cannot be extrapolated. Andrews et al. [11] consider the role of assumptions in modelling and interpreting simulation results.

Related patterns

There are assumptions made by the Domain Scientist about their domain before modelling even begins; Ask [Silly] Questions (217) to help tease these out, and beware of being an *Uncritical Domain Modeller* (143). Also, domain experiment compromises are different from simulation compromises; avoid being an *Uncritical Domain Scientist* (143) in accepting simulation compromises.

The Argument Modeller (106) ensures that assumptions are properly considered in order to Argue Instrument Fit For Purpose (186). Not all assumptions need go into the argument; by the end of the process, they have done their job of making the simulation fit-for-purpose.

Modelling Approach

Intent

Choose an appropriate modelling approach and notation.

Context

A component of the Domain Model (128) and of the Platform Model (149).

Discussion

The Domain Model is a rigorous descriptive (scientific) model of the Domain (123) of study. For building a Domain Model, choose an approach that is:

- understandable by the Domain Scientist (101), Domain Modeller (103), Simulation Engineer (105), Argument Modeller (106)
- suitable for expressing the Basic Domain Model (130) content and the Domain Behaviours (135)
- suitable for expressing the Simulation Behaviours (179) part of the Results Model (174)

The Platform Model (149) is a rigorous prescriptive (engineering) model of the Simulation Platform (161) implementation. For building a Platform Model, choose an approach that is:

- understandable by the Domain Modeller, Simulation Engineer, Argument Modeller

Choose a modelling approach with a well-defined notation, underpinned by a conceptual Metamodel (234) or a formal semantics. Choose a modelling approach capable of traceable mapping between models at different levels of abstraction (e.g. specification to design to implementation), ideally by supporting Seamless Development (214).

Where possible, use the same modelling approach for both models to support Seamless Development; however, as the models have different purposes and different audiences, there may not be a single modelling approach that is suitable for both models. If not, the argumentation between Domain Model and Platform Model will need more detail. Use a modelling approach for the Platform Model that supports Seamless Development to the Simulation Platform; it is better to have the complication of a change of approach at the higher level of models, than the lower level of moving from model to implementation.

Standard modelling approaches include Agent Based Modelling (208) (ABM), cellular automata (some authors class these as ABMs, but they are much more specialised), reactive systems, differential equation-based modelling, network-based models, stocks and flows systems models, Petri nets, and so on.

Modelling notations are the formal languages used to write down the model. The language used depends on the approach. For example, ODE models use differential calculus mathematical notations. Another approach is to use a rule-based modelling language, such as Kappa [35, 231].

Beware making up your own modelling approach: it is unlikely to have a well-defined semantics, making it essentially impossible to verify fitness-for-purpose. However, principled use of a well-defined Domain Specific Language (228) tailored for the Domain can enhance model comprehension. Webb and White [228] use UML for ABM of biological cells; Bauer and Odell [25] discuss the use of UML 2.0 for ABMs. Read et al. [187, 188] provide some guidance and discuss some limitations of UML for modelling certain aspects of biological systems.

Use your chosen modelling approach in a way that respects its defined semantics as much as possible. Record (Document Assumptions (108)) any variation from established semantics (e.g. the use of a class diagram to model agents not classes), and review its connotations (e.g. does the changed semantics affect the way that OO classes are implemented?).

Beware using an informal version of a language that does not respect its semantics.

Beware using an inappropriate notation that cannot express (or cannot express naturally) the model properties you want.

Beware using executable notations: they will include too much implementation detail, making it much harder to understand, and to argue fitness-for-purpose.

Related patterns

Examples of particular modelling approaches include Agent Based Modelling (208), and Environment Orientation (210).

Use a Hybrid Model (212) if different parts of the domain are naturally modelled using different approaches.

Use Seamless Development (214) where possible, but do not let it compromise building a natural domain model that is readily verifiable.

4.5 Antipatterns

Box Ticking

Problem

You are blindly following the CoSMoS approach in name only, as a bureaucratic box-ticking exercise. You are missing many of the benefits by not taking things seriously, or are doing too much work, by blindly following irrelevant patterns.

Context

The team has not bought in to the benefits of the CoSMoS approach, but the project is required to use it. Or the team is not thinking critically enough about the process.

Discussion

CoSMoS is a flexible approach, and needs to be tailored for each different project. It can be used in a "lightweight" manner for small projects, or in rigorous detail for larger critical projects. This tailoring requires buy-in from the team. If trying to use it on a large project, particularly for the first time, the details may seem burdensome, and so engagement drops, and "box ticking" starts. This can seriously impact the project, without providing any benefits.

Solution

Try building a smaller increment at first, to help appreciate the approach and its benefits.

Use the various arguments to decide whether you have enough detail, or whether a particular pattern is necessary at all.

Document the needed detail in your House Style (229). Think critically about whether a Partial Process (230) is suitable.

Cutting Corners

Problem

You follow the CoSMoS approach initially, but you do not revisit, update, or iterate.

Context

The team has used the CoSMoS approach initially, but is no longer using it.

Discussion

A symptom of *Cutting Corners* is out of date models, and inconsistencies between Domain Model, Platform Model, and Simulation Platform. This is a potential pitfall if development is rapid (see also *Premature Implementation* (170)), or if the value of the models is not appreciated.

Solution

Assign a Documentation Officer role. Ensure that the Project Repository (219) and Version Control (221) are being used. Document the needed steps in your House Style (229).

Divorced Argumentation

Problem

You do the argumentation as a separate exercise.

Context

The Argument Modeller (106) is divorced from the rest of the project team, and Argue Instrument Fit For Purpose (186) happens in isolation, or the role has not been assigned.

Discussion

You can argue fitness *retrospectively* if given a pre-existing simulation. In this case, the argument is necessarily built separately.

You can argue fitness *incrementally* during the development of a simulation. In this case, the argument should be built hand-in-hand with the development of the models and with the implementation and use of the platform. This helps the right level of detail to be modelled, and ensures that the necessary evidence is produced. If the argument is built in parallel, but as a separate exercise, these benefits are missed.

Solution

The Argument Modeller must actively collaborate with the rest of the team.

Chapter 5
Discovery phase patterns: building the domain model

Abstract In which we describe building an appropriate simulation platform — with patterns for the research context, the domain, and the domain model.

5.1 Catalogue of discovery phase patterns

Context setting patterns

Research Context (119)	Identify the overall scientific context and scope of the simulation-based research being conducted.
Simulation Purpose (121)	Agree the purpose for which the simulation is being built and used, within the research context.
Domain (123)	Identify the subject of simulation: the real world or engineered system and the relevant information known about it.
Cartoon (124)	Sketch an informal overview picture.
Expected Behaviours (126)	Describe the hypothesised behaviours and mechanisms.
Glossary (127)	Provide a common terminology across the simulation project.

© Springer Nature Switzerland AG 2018
S. Stepney, F.A.C. Polack, *Engineering Simulations as Scientific Instruments: A Pattern Language*, https://doi.org/10.1007/978-3-030-01938-9_5

Domain modelling patterns

Domain Model (128)	Produce an explicit description of the relevant domain concepts.
Basic Domain Model (130)	Build a detailed model of the basic domain concepts, components and processes.
Data Dictionary (132)	Define the data used to build the simulation and run experiments.
Domain Behaviours (135)	Describe the observed emergent behaviours of the underlying system.
Domain Experiment Model (135)	Define relevant experiments in the domain, as the basis for analogous simulation experiments and results analyses.

Antipatterns

Amateur Science (137)	You do not engage with a domain scientist, because you think you know the domain science well enough, or that the domain literature is sufficient input.
Analysis Paralysis (138)	You are spending too much time analysing and modelling the domain, trying to get everything perfect, and never getting to the simulation.
A Bespoke Too Far (139)	You invent a new modelling approach from scratch for your project.
Cartoon as Model (140)	Your domain model consists of nothing but cartoon sketches.
Executable Domain Model (140)	You write your domain model in an executable language, and use it as the simulation platform.
One Size Fits All (142)	You have a pre-determined modelling approach in mind.
Uncritical Domain Scientist (143)	The domain scientist accepts the domain model or the platform model on trust.
Uncritical Domain Modeller (143)	The domain modeller accepts everything the domain scientist says on trust.

5.2 Context setting patterns

Research Context

Intent

Identify the overall scientific context and scope of the simulation-based research being conducted.

Summary

- provide a brief *overview* of the research context
- document the *research goals* and project scope
- agree the Simulation Purpose (121), including criticality and impact
- identify the team members and their experience, and assign Roles (99)
- Document Assumptions (108) relevant to the research context
- note the available *resources*, timescales, and other constraints
- design and set up a Project Repository (219)
- determine *success criteria*
- decide whether to proceed, or walk away

Context

A component of the Discovery Phase (95), Development Phase (96), and Exploration Phase (97) patterns. Setting (and resetting) the scene for the whole simulation project.

Discussion

The role of the Research Context is to collate and track any contextual underpinnings of the simulation-based research, and the technical and human limitations (resources) of the work.

The Research Context comprises the high-level motivations or goals for the research use, the research questions to be addressed, hypotheses, general definitions, requirements for validation and evaluation, and success criteria (how will you know the simulation has been successful).

Consideration should be given to the intended criticality and impact of the simulation-based research. If these are judged to be high, then an exploration of how the work can be validated and evaluated should be carried out.

The scope of the research determines how the simulation results can be interpreted and applied. Importantly, it captures any requirements for validation and evaluation of simulation outputs. Assumptions may well constrain the project scope. Later assumptions at specific points in the project may imply a need to revisit the scope. It should also be revisited between phases, and potentially rescoped in the light of discovered knowledge.

Determine any constraints or requirements that apply to the project. These include the resources available (personnel and equipment), and the timescale for completion of each phase of the project. Any other constraints, such as necessity to publish results in a particular format (for example, using the ODD Protocol (222)), should be noted at this stage (and potentially added to the House Style (229)). This helps ensure that later design decisions do not violate the project constraints. Ensure that the research goals are achievable, given the constraints.

Limited resources, including staffing, competencies, data, and timescales, will limit the scope of the research. The project must lie within the team's Domain expertise, and within its modelling and implementation capabilities. Probing resourcing limits leads to discussion of factual and technical limitations, and helps to establish what is feasible, and to scope the simulation project appropriately.

Acknowledging limitations also exposes further assumptions. Probing of assumptions is intended to motivate discussion of the limitations of Domain knowledge and understanding, the appropriate Modelling Approach (111) and Implementation Approach (160), possible approaches to testing, and the approach to Argue Instrument Fit For Purpose (186).

Good software engineering practice is crucial for anything other than the simplest of simulations: a fit-for-purpose simulator to be used as a scientific instrument must be properly engineered. Prior software engineering expertise is useful, not least in avoiding well-known development pitfalls, but it is possible to develop a demonstrably fit-for-purpose simulation using good programming skills and practices, and following good practice guidelines such as the CoSMoS approach.

Decide relevant success criteria. What would constitute a successful project? Are you expecting or hoping for failure (that is, to discover model inadequacies), or expecting confirmation of a hypothesised mechanism? Such confirmation will require good argumentation

Having established a realistic scope for the project, now is the time to decide whether to proceed, or walk away. The constraints (competencies, timescales, etc.) may have limited scope so much that it is too narrow, and the project not worthwhile.

Related patterns

The research context scopes what should go in the models and simulation: it can help avoid *Analysis Paralysis* (138).

It is important to identify if, when, why and how the Research Context changes throughout the course of developing and using the Simulation Platform (161).

Simulation Purpose

Intent

Agree the purpose for which the simulation is being built and used, within the research context.

Summary

- define the *role of the simulation*
- determine the *criticality of the simulation results*

Context

A component of the Research Context (119).

Discussion

The simulation project has a *purpose*, a role to play, within the overall Research Context (119). The purpose of a simulation exercise is the single most important concept. Without a defined purpose, it is impossible to scope the Research Context, and it is impossible to arrive at a consensus over fitness for purpose.

The purpose might be a proof of concept, to explore a mechanism, to suggest new experiments, to test a hypothesis, to predict an outcome, to educate, or more [65, 83]. The purpose needs to be made explicit, as it forms the basis of the fitness for purpose arguments. A simulation might be fit for an exploratory purpose, but not rigorously enough developed for a predictive purpose; it might be suitable only as a Prototype (213). Alternatively, a simulation built for a predictive purpose might be expensively overengineered for a simpler exploratory purpose.

The purpose determines the appropriate levels of abstraction for modelling, the appropriate implementation languages and platform, and the appropriate analysis and interpretation of results.

Purpose is closely tied to criticality and impact. If the purpose is to provide first-class evidence for major research, then the development approach should exploit state-of-the-art software engineering methods, validation, and documentation approaches; the simulator and its fitness-for-purpose must be capable of international expert scrutiny. An exploration of how the work can be validated and evaluated should be carried out.

In the more usual case, the purpose is to provide a test-bed for hunches and a generator of hypotheses, all of which will then be subject to conventional experimental analysis and confirmation before publication. Here the development and argumentation need to be just good enough: internal consensus and internal documentation are sufficient.

A key feature of *purpose* is its role in scoping. A Simulation Platform (161) that is designed for one purpose may or may not be modifiable for another purpose. Ideally, we want flexible, extensible platforms, but this should not be at the expense of quality. For example, our running example of the prostate cell differentiation and division model (Chapter 3) has the explicit purpose of simulating cancer neogenesis. Based on the particular Domain Scientist view of cancer neogenesis, we break this into two purposes, allowing the phasing of the simulator development: (a) capturing "normal" prostate cell differentiation and division; (b) expressing biologically-realistic mutability and heritability. Note that the purpose is firmly rooted in the Domain, and in the needs of the Domain Scientist; it is not a computational goal.

Related patterns

It is important to identify if, when, why and how the simulation purpose changes throughout the course of Multi-increment Simulation (240) project. Revisit the Research Context (119) and Simulation Purpose (121), and associated models and arguments.

Domain

Intent

Identify the subject of simulation: the real world or engineered system and the relevant information known about it.

Summary

- draw an explanatory Cartoon (124) of the domain
- provide an overview description of the domain
- define the Expected Behaviours (126)
- provide a Glossary (127) of relevant domain-specific terminology
- Document Assumptions (108) relevant to the domain
- define the scope and boundary of the domain – what is inside and what is outside – from the Research Context (119)
- identify relevant sources: people, literature, data, models, etc.

Context

A component of the Discovery Phase (95). The Research Context (119) sets the boundaries of the domain: what is within and out of scope.

Discussion

The Domain is the part of the real world of interest, the part that the simulation project is "about". Domain knowledge is "owned" by the Domain Scientist (101), who provides information and approves the form and content of the captured knowledge.

The general domain will have been identified by the Research Context (119). This pattern refines that identification, by identifying the required source of the information that will inform the Domain Model (128), and relevant underlying assumptions.

No formal modelling takes place at this stage: that is the job of the Domain Model (128). However, an informal Cartoon (124) and descriptions help set the scene and provide links into the literature, particularly with respect to defining the scope and boundary: what will be modelled, and what will not.

Related patterns

For a Multi-increment Simulation (240), the scope and boundary define the current increment, and will potentially be expanded in later increments.

If the domain is an Engineered Domain (238), rather than a natural domain, the CoSMoS approach can still be used, but the results need to be interpreted differently.

Cartoon

Intent

Sketch an informal overview picture.

Context

A component of the Domain (123) and Domain Model (128).

Discussion

The Cartoon captures the important aspects of the Domain or Domain Model. It gives an initial view of the relevant components and behaviours that need to be captured, in informal, domain-specific terms. It provides the initial view of the scope and level of detail.

We use the term "cartoon" for any figure that has informal components[1]. This may be due to it simply being a sketch, or because various arrows and boxes are used in ill-defined or ambiguous ways.

A Cartoon can be an abstract sketch of the spatial properties of the system, such as the layout of prostate cells (Figure 3.1, p.55). It can be a sketch map of terrain. A Cartoon may be provided by the Domain Scientist (101) to help explain the Domain; it might be drawn by the Domain Modeller to help express their understanding of the Domain. In either case, the Domain Scientist should confirm that the Cartoon is useful, and is a suitable representation of the knowledge being captured.

A Cartoon can be thought of as a proto-model, with no (explicit) Metamodel (234), and no (explicit) semantics or even syntax. A Cartoon is more analogous to an architect's conceptual sketch, whereas the Domain Model is more analogous to the architectural plans; continuing this analogy, the Platform Model is analogous to the builders' plans. Other places where cartoons or sketches are used as part of a design process include [42].

Do not spend effort "prettifying" a Cartoon. It should be sketchy enough to be obviously informal, and for it to be clear that more information than shown is needed to build the Simulation Platform. It is for communication between people, so it should be laid out for understanding, not just splattered down at random on the page.

A spectrum of Cartoons of increasing abstraction and formalism can help provide a bridge between the Domain and the Domain Model itself. For example, compare the Domain cartoon of Figure 3.1, p.55 with the associated Domain Model cartoons of Figures 3.3 and 3.4, p.62.

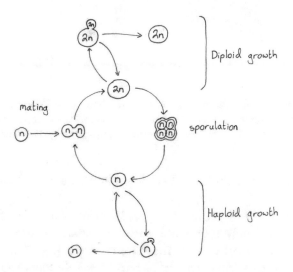

Fig. 5.1 A typical biological Cartoon, of the yeast life cycle. It is clear, given accompanying explanation, what this means. Parts of this cartoon have superficial resemblance to a state diagram, parts to a Petri net. But there are ambiguities. For example, in the "haploid growth" cycle, two arrows leave the "budding n" state: this is cell division, both arrows are taken. Yet in the main cycle, two arrows leave the "2n" state: this is a behaviour "choice", and one or other arrow is taken. Different cartoons may use similar arrows with yet further conventions: every cartoon has its own specific diagrammatic "semantics"

A Cartoon may be a sketch of the dynamics of the system, such as the lifecycle (Figure 5.1).

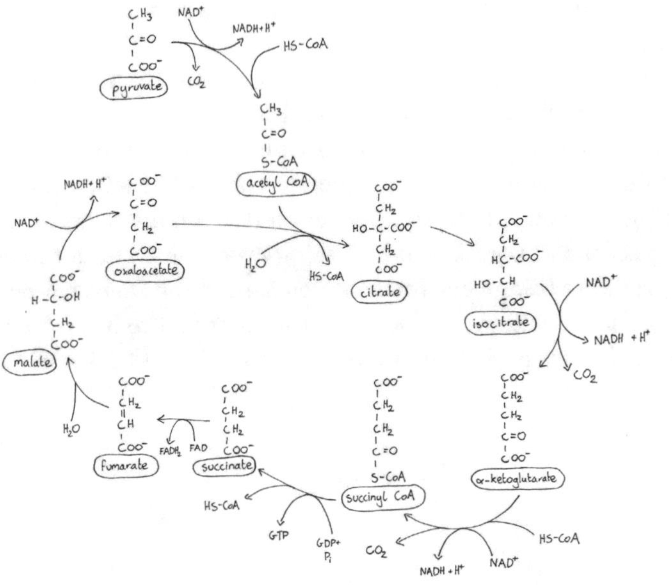

Fig. 5.2 A biological cartoon, of the citric acid (Krebs, or TCA) cycle, a metabolic pathway. The notation here is conventional, with different arcs representing the reaction in the main cycle, and showing the incoming and outgoing components

A Cartoon may include feedback loops and other causal arrows (Figure 5.2).

Any biology textbook, for example [2], makes heavy use of such Cartoons. Systems Analysis also makes use of feedback pictures, sketching the stocks and flows in a system, see for example, [72, 172].

Checkland uses 'rich pictures' as part of Soft Systems Methodology [50, 51]. A rich picture is a Cartoon that provides a conceptual map of the objects of interest and the relationships between them. Examples of using rich pictures during the process of object oriented systems modelling can be found in [150].

Expected Behaviours

Intent

Describe the hypothesised behaviours and mechanisms.

Context

A component of the Domain (123), in the context of the given Research Context (119).

Discussion

The Expected Behaviours is a description that encapsulates a summary of what is observable in the Domain, what concepts are probably involved and how, and what hypotheses related to these observables we want to investigate.

The Expected Behaviours includes documentation of the hypotheses related to the Domain that are to be investigated and tested by simulation. It captures the system level, or global, emergent properties of interest in the Domain, and the local mechanisms hypothesised to give rise to these global properties.

It can take the place of some traditional requirements analysis, capturing what aspects of the domain are to be modelled, and how to link them back to the domain. See, for example, Figure 3.2. So it provides a focus for determining what should be observable in the simulator's Results Model (174), either directly or through a surrogate.

This deceptively simple cartoon can take much effort to construct by the Domain Scientist (101) and Domain Modeller (103), as they unpick what is known about the observable Domain, and what is measurable.

Related patterns

The relevant expected behaviours should be observable through measurements captured in the Domain Experiment Model (135).

The hypotheses will be investigated in a Simulation Experiment (177).

Glossary

Intent

Provide a common terminology across the simulation project.

Summary

- provide a list of key terms and standard abbreviations for use in the project.

Context

A component of the Domain (123) description.

Discussion

It is important to agree a common vocabulary (and its meaning) for use in a project. Good names are crucial to communication. Developers have to invent names in their models and code. Where possible, they should use ones meaningful to the domain scientists, to aid communication. A glossary provides the source of such names.

Each name should have a short definition, and may be accompanied by a standard abbreviation. Note when alternative names are used in the literature.

Add key terms to the glossary as they become important in the project: do not try to build a list of the entire domain vocabulary.

Beware of these names changing from the Domain meaning on the move from the Domain to the Platform Model. In particular, do not abstract several domain types into a single platform type and give it the name of one component type: do not abstract the combination X, Y, Z as X. Also, do not split one domain type into platform subtypes and give one of the subtypes the name of the main type: do not split X into X, Y, Z.

Related patterns

If the glossary gets inconsistent, Refactor (233) it, and its use in project documentation.

5.3 Domain modelling patterns

Domain Model

Intent

Produce an explicit description of the relevant domain concepts.

Summary

- draw an explanatory Cartoon (124)
- discuss and choose the domain Modelling Approach (111) and level of abstraction
- define the Domain Behaviours (135)
- build the Basic Domain Model (130) using the chosen modelling approach
- build the Domain Experiment Model (135)
- build the Data Dictionary (132)
- build the domain Stochasticity Model (154)
- Document Assumptions (108) relevant to the domain model

Context

A component of the Discovery Phase (95), in the context of a given Domain (123) and Expected Behaviours (126) description.

Discussion

The Domain Model is an explicit model of the agreed scientific basis for the development of a Simulation Platform (161)[2]. It is a descriptive (scientific) model of the Domain, as understood from domain experiments, observations, and hypotheses of underlying mechanisms.

A model (domain or otherwise) can have several components, each expressed in some notation: text, Cartoon, formal diagram, mathematics, pseudocode, code. Choose the most appropriate notation(s) for each component. Criteria include naturalness of the form, level of abstraction, required precision, and comprehensibility to intended readership.

The Domain Model is more formal than the description of the Domain, and potentially of more limited scope, since the Domain may have contextual information not captured here. A new Cartoon can help highlight this change of formality and scope.

The Domain Model itself has three main component (see Figure 1.3, p.21):

1. a Domain Experiment Model (135), capturing the experimental protocols
2. a Basic Domain Model (130), capturing the low level core components, structures, interactions and behaviours of the system

3. a Domain Behaviours (135) model, capturing the high level properties observed of the system.

These models rest on a common Data Dictionary (132).

The domain model should as far as possible be *platform neutral*. Domain model simplifications and abstractions must make sense in *domain* terms, and not be made purely for implementation reasons. Any further simplifications and assumptions necessary for implementation should be made on moving to the Platform Model.

Any domain will have some degree of stochasticity, due to noise, uncertainties, small samples, or other effects. This should be captured in a Stochasticity Model (154), enabling the simulation of stochasticity to be in domain terms.

Collaboration between the Domain Modeller (103), Argument Modeller (106), and Domain Scientist (101) is essential when working in cross-disciplinary science. Building fit-for-purpose models requires expertise from both the specific domain and the modelling process. This level of interconnection of skill is greatly helped by face-to-face collaboration where free discussion to take place. Questioning should be actively encouraged, as many implicit assumptions can be identified by asking for the rationale underlying various statements.

Related patterns

If the domain is an Engineered Domain (238), its domain model is instead a prescriptive (engineering) model. However, it is still a distinct model from the Platform Model.

Basic Domain Model

Intent

Build a detailed model of the basic low level domain concepts, components and processes.

Context

A component of the Domain Model (128), in the context of a given Domain (123), built by the Domain Modeller (103), using the chosen domain Modelling Approach (111).

Discussion

The Basic Domain Model should capture the low level components and processes, using the domain Modelling Approach. These are the concepts that will be transferred into the Platform Model and Simulation Platform, albeit with some modifications. These concepts should be captured in sufficient detail for subsequent simulation, but in Domain terms, such that the model can be agreed by the Domain Scientist.

For example, in the running example in Chapter 3, the Basic Domain Model comprises the Petri net and state machine models of the various cell division and differentiation processes. The four main cell types were supplied by the Domain Scientist. An auxiliary type, the 'daughter cell', was introduced by the Domain Modeller to simplify and clarify the model. This simplification was discussed in detail with the Domain Scientist, to ensure that it was acceptable to them.

For example, in the CellBranch case study in part IV, the Basic Domain Model is presented as two models, of increasing abstraction. The 'Transcription Factor Branching Process' description is already an abstraction away from the actual processes occurring in transcription, developed by the Domain Scientist: testing the adequacy of this abstraction is the major part of the Research Context. This model is further abstracted to the 'sparking posts' model, by the Domain Modeller, for use as the basis for further development. Again, this final model was discussed in detail with the Domain Scientist, and their approval of the abstract gained. The model is formalised in a UML class diagram and state diagram.

Related patterns

The Domain Experiment Model (135) captures concepts of the experimental setup, and the Domain Behaviours (135) model captures high level, or emergent properties, so these concepts should not be included in the Basic Domain Model. Parameters and variables are captured in the accompanying Data Dictionary (132).

Data Dictionary

Intent

Define the modelling data used to build the simulation, and the experimental data that is produced by domain experiments and corresponding simulation experiments.

Summary

- define and capture the modelling data used to build the simulation
- define and capture the experimental data that provides the comparison between the Domain Model (128) and the Results Model (174)
- determine whether there is sufficient domain experimental data to provide Calibration (163), validation, and unseen acceptance test data sets

Context

A component of the Domain Model (128), Platform Model (149), and Results Model (174).

The Data Dictionary links the various data sources and results in the Domain Experiment Model (135), the Simulation Experiment Model (152), and the Simulation Experiment (177).

There is observational data that is present in the Domain Model. It needs to have instrumentation provided for in the Platform Model and the Simulation Platform, to extract the analogous data from the simulation. This model is also used to capture the simulation outputs as part of the Results Model.

Discussion

The Domain includes identification of the data sources that populate the Data Dictionary. There are two kinds of data considered:

1. *modelling data (parameter values)*: used to parameterise the various models, by providing numbers, sizes, timescales, rates, and other system-specific values; this usually comes from the raw data from previous experiments, analysed and reduced using previous models and theories

2. *experimental data*: comprising the input values and output results of the domain experiments and corresponding simulation experiments; this is broken into three parts:

   a. Calibration (163) data, for tuning the platform parameter values
   b. validation data, to allow the calibrated Simulation Platform platform to be validated against the Domain Model[3].
   c. further unseen data, for comparing with Simulation Experiments

In some systems there may be insufficient experimental data to perform calibration and validation. If so, an argument should be used to demonstrate why this is not considered to be a problem.

If the simulation has high criticality (determined from the Research Context (119)), it would be reasonable to require a further set of truly unseen validation data, to form the basis for an "acceptance test", before the system is used in any critical capacity.

Domain Model parameter and data values might be directly used in the Platform Model and Simulation Platform. For example, environmental parameters such as rainfall rates in an ecological simulation, or robot sensor data in an engineering simulation, might carry over unchanged. However, Domain Model parameter and data values are not necessarily identical to their Platform Model counterparts. For example, a single value represented in the Simulation Platform might be a proxy for a collection of values in the Domain. So there needs to be a well-defined translation that maps these values between models, captured by the "informs" arrow in Figure 5.3. Similarly, the data output from a Simulation Experiment, captured and analysed in the Results Model, needs to be translated into Domain Model terms, to allow comparison. The form of these translations is guided by the translation of domain model concepts to platform model concepts, and the precise structure is determined by Calibration.

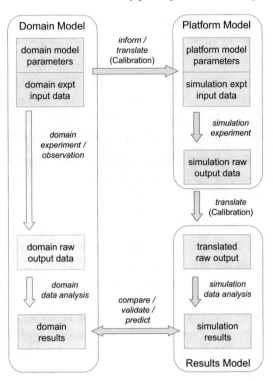

Fig. 5.3 The components of the data dictionary in the Domain Model, and how they relate to components in the Platform Model and the Results Model. The specific translations are established during Calibration

It is possible to extract much more information from a Simulation Experiment than from a domain experiment, but if it is not observable (even indirectly, through surrogates, or by investigating predictions) in the Domain, it is of little use.

The necessity for suitable data in the Results Model implies requirements on the Platform Model: it must be of a form that can produce the required data, and must be suitably instrumented to output the data.

This careful separation of modelling data (used to build the model) and experimental data (to be produced by the domain experiment or analogous Simulation Experiment) is important, in order not to *Program In the Answer* (170).

Related patterns

A Visualisation Model (181) can be used for presenting Simulation Experiment output data to the user in Domain terms.

Domain Behaviours

Intent

Describe the observed emergent behaviours of the underlying system.

Context

A component of the Domain Model (128), in the context of a given Domain (123), built by the Domain Modeller (103).

Discussion

The Domain Behaviours is the model the emergent system level, or global, properties and behaviours of the Domain. It formalises the emergent behaviours captured in the Expected Behaviours (126) model. It provides the link between the Domain Model and how the simulation Results Model should to relate to it.

Related patterns

The model has a simulated analogue in the Results Model (174). The relevant domain behaviours should be observable through measurements captured in the Domain Experiment Model (135), and their analogues in the Simulation Experiment (177) model.

Domain Experiment Model

Intent

Define relevant experiments in the Domain, as the basis for analogous Simulation Experiments and results analyses.

Context

A component of the Domain Model (128), in the context of a given Domain (123), built by the Domain Modeller (103).

Discussion

The Domain Experiment Model captures how the concepts, structures and behaviours in the Domain Model are controlled and manipulated, such as in wet lab experiments or measurements in the field.

It details what data is collected from the experiments and how that data is then manipulated and interpreted, for example, using statistical methods, to produce the experimental results. This provides the link to the Results Model (174), which provides analogous measurements and manipulations, to allow the domain and simulation outputs to be meaningfully compared.

An example of where the structure of domain experiments is mirrored in a simulator is Aevol, an *in silico* experimental *artificial evolution* platform [24], which encapsulates an *in silico* laboratory to test evolutionary scenarios mimicking those used in real bacterial evolutionary studies[4].

The assessment of Simulation Experiment results in terms of system-level emergent domain properties can pose a challenge to the corresponding Simulation Experiments. The simulated domain typically captures only a subset of the experimental domain system, of which no corresponding observation can be made. For example, the domain observation might be on a six point scale from "no symptoms" to "death" of an experimental organism, whereas the simulation might be at the tissue level, complicating both Calibration (163) and eventual comparison between simulation and domain experiments[5].

Related patterns

The specific details of the experimental data and parameters are captured in the Data Dictionary (132). Relevant details of the Domain Experiment Model are carried over to the Simulation Experiment (177) model. The data and analyses have simulated analogues in the Results Model (174).

5.4 Antipatterns

Amateur Science

Problem

The Domain Modeller does not engage with a Domain Scientist, because they think they know the Domain science well enough, or that the Domain literature is sufficient input.

Context

Building the Domain Model (128); making simplifying assumptions in the Platform Model (149); performing platform Calibration (163); building the Results Model (174); running a Simulation Experiment (177).

Discussion

While modelling it can be easy for the Domain Modeller to use their own understanding of the Domain rather than referring to the Domain Scientist (101). This understanding is, however, nearly always oversimplified and at too shallow a level: even if a Domain looks relatively straightforward from the outside, it can have hidden subtleties and surprises. After all, if it really were that simple, there would be no need for a simulation instrument.

The Domain Modeller may assume that the scientific literature is complete, consistent, pertinent, and correct, and that it is sufficient to build the Domain Model. However, technical scientific literature relies on much tacit knowledge for understanding. The literature is vast, and the Domain Scientist can help filter the material that is relevant for the Simulation Purpose (121). The literature contains terminology clashes, with sometimes different names referring to the same concept, and worse, the same name referring to different concepts. It contains errors. The Domain Scientist can help the Domain Modeller navigate this morass.

If you are finding it difficult to Document Assumptions (108) about the Domain or Domain Model, you may be engaged in *Amateur Science*.

Solution

Engage with the Domain Scientist, who will soon make it clear that the real world Domain "is more complicated than that". They can advise which literature is key, which is peripheral, which is mistaken, and can provide the tacit domain wisdom not present in any literature. Use the Glossary (127) to record terminology clashes and ambiguities.

But beware being an *Uncritical Domain Modeller* (143) placing blind trust in the Domain Scientist.

Analysis Paralysis

Problem

You are spending too much time analysing and modelling the Domain, trying to get everything perfect, and never getting to the implementation.

Context

Building the Domain Model (128), the Platform Model (149), or the Results Model (174).

Discussion

Analysis Paralysis is a classic problem in many areas where a problem needs to be analysed and modelled before it can be solved. It can be due to a combination of wanting to get it "right first time", and a fear of implementation failure.

One symptom is to Refactor (233) over and over, to "improve" the model's structure or genericity, without making progress.

Another symptom is putting irrelevant information or detail into a model, just because you can. This leads to large, clumsy, buggy, incomprehensible models that are hard to develop, simulate, analyse, and understand. Even if the model "works", it may provide no more insight into the problem than the actual (complicated) Domain does. The key to good models is finding a suitable level of abstraction, and suppression of irrelevant detail. Everything in the model needs "to earn its keep".

Solution

Keep the Research Context (119) constantly in mind: do not analyse, model or debate anything that does not serve that context. If the context requires different levels of detail, consider building a Multi-scale Simulation (244).

Do not fear making a mistake; recognise that you will not get it "right first time". Instead, Prototype (213) to explore possible approaches, and use Multi-increment Simulation (240) to focus on smaller parts of the problem. There are some techniques for simplifying simulations, such as Grimm et al.'s "pattern-oriented modelling" [106], and the method of Van Nes and Scheffer [165].

Beware also the opposite anti-pattern: *Premature Implementation* (170).

A Bespoke Too Far

Problem

You invent a new modelling approach from scratch for your project.

Context

Building the Domain Model (128).

Discussion

Very often, none of the modelling approaches that exist will perfectly fit what you want to capture in your Domain Model. Since the Domain Model is for communication with the Domain Scientist, not for direct execution, you decide to invent an appropriate modelling approach from scratch.

Inventing sound modelling approaches is difficult. Such an approach will either take up all or more of your project resources, or, more likely, will result in a modelling approach that is ill-defined and ambiguous, leading to difficultly in transforming your Domain Model to a suitable Platform Model (149).

Solution

Instead, you might consider developing a Domain Specific Language (228) that interfaces to a well-defined modelling language.

Beware also the opposite anti-pattern: *One Size Fits All* (142).

Cartoon as Model

Problem

Your domain model consists of nothing but Cartoon sketches.

Context

Building the Domain Model (128).

Discussion

A Cartoon (124) provides a good overview or summary of a model, but it is not sufficient as a model itself. It lacks detail, and specificity: some of the properties that make it a good overview. A full Domain Model needs more information, and needs to be more formally defined, in order to form a sound basis for transformation to a Platform Model (149).

Solution

Use the Cartoon as an overview, that accompanies a full Domain Model. Explicitly label it "cartoon" in order to emphasise that it is not the model itself. But avoid the *Executable Domain Model* (140).

Executable Domain Model

Problem

You write your Domain Model in an executable language, and use it as the Simulation Platform.

Context

Building the Domain Model (128).

Discussion

You have to write your Domain Model in some notation, so why not chose an executable one, and save all the hassle of having to do more work to implement a Simulation Platform? (There are practitioners who advocate this approach.)

There are several problems with such an approach if you are engineering a simulation for use as a scientific instrument.

- The Domain Model is a scientific (descriptive) model of the Domain; the Platform Model is an engineering (prescriptive) model of the *simulation* of the Domain. Although they overlap to some degree, they address different concerns.
- An *Executable Domain Model* will be cluttered with implementation details, making it hard for the Domain Scientist (101) to understand, and hard to Argue Instrument Fit For Purpose (186).
- An *Executable Domain Model* may contain Domain details that should not be in the Platform Model or Simulation Platform, such as emergent Domain Behaviours (135), or other hypothesised results.
- Alternatively, an *Executable Domain Model* may comprise only the Basic Domain Model (130), and omit the definition of the Domain Behaviours, forestalling any rigorous validation.
- It is very easy to start "hacking" an executable model, making it harder to understand, and to forget the necessary accompanying documentation of assumptions and simplifications that allow you to Argue Instrument Fit For Purpose (186).

Solution

Model in Domain terms, in a platform neutral manner. Explicitly split the model into its components (Basic Domain Model (130), Domain Behaviours (135), Domain Experiment Model (135)), so that emergent properties are captured.

In some cases, using an appropriate high level Domain Specific Language (228), it may be possible to express some or all of the Domain Model in an executable language at a suitable level of abstraction. In such a case, examine carefully whether this level of detail and precision is necessary for the Research Context (119) and Simulation Purpose (121). If done, rigorously avoid adding detail that is only for implementation purposes. The relevant Platform Model should then be derived from the Basic Domain Model and Domain Experiment Model following the same process as for a non-executable model. This is an extreme example of Seamless Development (214).

Beware the opposite extreme: *Cartoon as Model* (140).

One Size Fits All

Problem

You have a pre-determined Modelling Approach in mind. Irrespective of the specific Research Context or scientific question, you build the Domain Model using this given modelling approach, and disregard its possible disadvantages.

Context

Discovery Phase (95), Development Phase (96)

Discussion

A Domain Modeller (103) may become proficient in using a certain Modelling Approach (111). Being expert in defining models in a given paradigm (for example, Agent Based Modelling (208), or ODEs), their interaction with the Domain Scientist (101) is structured according to that paradigm's concepts.

The team will potentially fail to acknowledge the limitations of their approach; the project may overrun its agreed resources. Using an inappropriate Modelling Approach may lead towards complicated, costly simulators, or the problems may not be solvable in the given time.

Similar comments apply to using a pre-determined Implementation Approach (160): beware of using an implementation framework simply because it is familiar. And beware of biassing the choice of Modelling Approach in order to target a such a framework.

Solution

Ideally, choose an appropriate Modelling Approach based on the Research Context (119). Ensure that any pre-determined choice in your House Style (229) is properly scrutinised.

However, the modelling expertise may be a fixed component of the project team. If so, the Domain Modeller should inform the Domain Scientist of the approach's advantages and limitations. Because the Modelling Approach influences the type of data that is required (e.g. for building bottom-up ABMs, data describing individual entities is more important than population-level data), the team must identify if such data can be obtained. If not, the Research Context may need to be redefined.

Beware also the opposite anti-pattern: *A Bespoke Too Far* (139).

Uncritical Domain Scientist

Problem

The Domain Scientist accepts the Domain Model or the Platform Model on trust.

Context

Discovery Phase (95); Argue Instrument Fit For Purpose (186): reviewing arguments to ensure consensus

Discussion

A Domain Scientist (101) should not have to understand software engineering, but uncritical acceptance of a computer simulation may mean that misinterpretations or inappropriate assumptions of the Domain (123) and Domain Model (128) are not identified. This may be detected during Calibration (163) or in construction of the Results Model (174). In the worst case, uncritical acceptance of the simulation leads to the Domain Scientist making inappropriate inference from the simulation results.

An *Uncritical Domain Scientist* may arise because there is insufficient buy-in to the collaboration, or because each expert is too inclined to accept at face value the expertise of others.

Solution

Nominate a person within the Domain Scientist role who is responsible for challenging the modelling and coding decisions of the Domain Modeller. The person must feel empowered to Ask [Silly] Questions (217).

Uncritical Domain Modeller

Problem

The Domain Modeller accepts everything the Domain Scientist says on trust.

Context

Discovery Phase (95); Argue Instrument Fit For Purpose (186): reviewing arguments to ensure consensus

Discussion

The Domain Modeller (103) should not have to understand all the background of the Domain (123): it is the role of the Domain Scientist (101) to select and interpret the Domain for the Domain Modeller. However, the Domain Modeller must challenge and probe the understanding of the Domain Scientist in order to uncover meanings, limitations, uncertainties etc.

A common situation that is revealed by probing is the Domain "common knowledge" problem: something that was introduced to address a problem many years ago, and has been used habitually since, but was never justified or rationalised. A "why" question might reveal that the Domain Scientist uses the "something" because that is "the established practice": it may be a perfectly good fix, but equally it may be hiding or surrogating a real issue.

Another common situation is where the Domain Scientist makes a confident assertion about the structure of a complex system, or the results of a well-known set of experiments. Probing questions can reveal uncertainties, missing links, data that is collected from many (different) subjects, data point values that would be better stated as ranges of possible values, etc.

Sometimes the Domain Scientist will make tacit assumptions based on their own belief about how the simulation will work. They may make unnecessary and unwarranted simplifications, to help "improve efficiency". Alternatively, they may try to insist on too much detail, because they fail to see simplifications and generalisations that the Domain Modeller can make.

Solution

Nominate a person within the Domain Modeller role who is responsible for challenging the input of the Domain Scientist. The person must feel empowered to Ask [Silly] Questions (217); they should understand the reason and justification for all assumption and details, and should Document Assumptions (108).

It is important to understand the basis on which any categorical statement is made, and an argument can be used to record the source of facts, assumptions, justifications etc. Construction of the argument can be used to motivate the search for sources of domain knowledge etc.

Also avoid *Amateur Science* (137).

Chapter 6
Development phase patterns: developing the platform

Abstract — In which we describe developing the simulation platform appropriately, with patterns for the platform model and the simulation platform

6.1 Catalogue of development phase patterns

Platform Modelling patterns

Platform Model (149)	From the Domain Model, develop a platform model suitable to form the requirements specification for the Simulation Platform.
Basic Platform Model (151)	Build a detailed model of the basic platform concepts, components and processes.
Simulation Experiment Model (152)	Define relevant experiments in the simulation, analogous to domain experiments.
Stochasticity Model (154)	Model any required stochasticities explicitly.
Space Model (155)	Define how physical space is modelled.
Time Model (158)	Define how physical time is modelled.

© Springer Nature Switzerland AG 2018
S. Stepney, F.A.C. Polack, *Engineering Simulations as Scientific Instruments: A Pattern Language*, https://doi.org/10.1007/978-3-030-01938-9_6

Implementation patterns

Implementation Approach (160)	Choose an appropriate implementation approach and language.
Simulation Platform (161)	Develop the executable simulation platform that can be used to run the Simulation Experiment.
Calibration (163)	Tune the Simulation Platform so that simulation results match the calibration data provided in the Data Dictionary.
Debug by Video (167)	Use visualisation to help debug the simulation platform.
Propagate Changes (168)	Ensure that changes in one part of the system propagate throughout, to ensure consistency.

Antipatterns

Amateur Coding (169)	You believe that "anyone can write code".
Premature Implementation (170)	You start writing Simulation Platform code before having a proper understanding of the domain or platform requirements.
Program In the Answer (170)	The results from the simulation are an inevitable consequence of the simulation programming, not an emergent consequence of the operation of the simulation.
Independent Simulation Engineer (171)	The Simulation Engineer diverges from the Domain Model in order to create a more aesthetic or efficient computer program.

6.2 Platform modelling patterns

There is considerable traditional software engineering involved in developing the platform model and simulation platform. We do not cover all of that here here; we cover only the parts that are CoSMoS specific.

Platform Model

Intent

From the Domain Model (128), develop a platform model suitable to form the requirements specification for the Simulation Platform (161).

Summary

- choose a Modelling Approach (111) and application architecture for the platform modelling
- develop the platform model from the Domain Model (128). In particular:

  - remove the Domain Behaviours (135)
  - develop the Basic Platform Model (151) from the Basic Domain Model (130)
  - develop the Simulation Experiment Model (152) from the Domain Experiment Model (135)

- Document Assumptions (108) relevant to the platform model
- if necessary, Propagate Changes (168)

Context

A component of the Development Phase (96), with an existing Domain Model (128).

Discussion

The Platform Model is a prescriptive (engineering) model, and provides the high level specification of the Simulation Platform. It is developed from the Domain Model. Whereas the Domain Model is a model of the real world Domain, the Platform Model is a model of a software *simulation* of that Domain.

The chosen Modelling Approach should be of a form, and use a notation suitable for, proceeding to an implementation. This choice should be made in concert with choosing an Implementation Approach for the Simulation Platform (161), since they are so closely linked, and because some of the design decisions made in building the Basic Platform Model may be contingent on the eventual implementation route. However, the Platform Model should be kept as free from implementation details as is practical.

As far as possible, this should follow the principle of Seamless Development (214) and not introduce an unnecessary significant change from the chosen Domain Model approach. However, as noted above, the purposes of the two models are different: the Domain Model captures the Domain in terms comprehensible to the Domain Scientist (101); the Platform Model captures the specification of the simulator in terms and at a level of precision suitable for development. As such, different modelling approaches and notations may be needed. In such a case, there will need to be more argumentation of how the Platform Model correctly captures the Domain Model.

This is the point to determine the application architecture for the Simulation Platform, as this will affect the structure of the Platform Model.

The Platform Model can be developed from the Domain Model in the following manner:

- It is important *not* to carry over aspects of the Domain Behaviours (135) model that capture the "answer", or hypotheses to be investigated, emergent properties, etc.
- The Basic Platform Model (151) should be developed from the Basic Domain Model (130).
- The Simulation Experiment Model (152) should be developed from the Domain Experiment Model (135).

In general, given a hypothesis under consideration, components in the Domain Model that are *outcomes* of hypothesised mechanisms should not appear in the Platform Model: the answer should not be explicitly coded into the Simulation Platform; but it needs to appear in some model, for validation purposes.

During the process of developing the Platform Model, ambiguities, errors or inadequacies may be discovered in the Domain Model. If so, it is important to Propagate Changes through all the models and arguments to ensure that the whole development is consistent.

Basic Platform Model

Intent

Build a detailed model of the basic platform concepts, components and processes.

Summary

- develop the Basic Platform Model (151) from the Basic Domain Model (130)
- as needed, develop the Stochasticity Model (154)
- as needed, develop the Space Model (155)
- as needed, develop the Time Model (158)

Context

Part of the Platform Model (149); the computational realisation of the Basic Domain Model (130).

Discussion

The Basic Platform Model captures the computational realisation of the domain micro structures and behaviours, in a form suitable for implementation in software. This moves from a model of the real world micro-domain, to a model of the simulation of components of the micro-domain. This move requires a translation of concepts between the chosen Modelling Approach (111) for the Domain Model and for the Platform Model. Seamless Development (214) reduces, but does not obviate, the need for such a translation; even if the modelling approach is common between models, the models are of different entities: domain entities *versus* simulation entities. This translation process has several aspects:

- Determine a systematic conversion between the Domain Model language and the Platform Model language, paying attention to the Glossary and record of assumptions and justifications. For the Prostate example in Chapter 3, this involves a translation from a Petri net to a UML model.
- Where the Domain Model components do not map directly to Platform Model language concepts (for example, agents or classes or processes), determine a suitable construction of appropriate imple-

mentation structures (for example, state diagrams implemented by
a suite of classes), keeping a record of the traceability, and of any
assumptions that need to be made.
- Some components may be changed from Domain Model to Platform
  Model: there may be a need for further simplifications, proxies, surrogates, and so on. Determine the needed changes, and the necessary additions to the Data Dictionary (132), and document the relationships, for use in Calibration (163).
- Walk through the resulting Platform Model to check that it is syn-
  tactically and semantically consistent (for example, where transitions are guarded with Boolean expressions, there is a transition
  for every possible condition).

Further components need to be added to the Basic Platform Model,
particularly to simulate features of the Domain taken as "given": these
include the platform Stochasticity Model (154), Space Model (155) and
Time Model (158), any physical properties such as movement, gravity,
temperature, that are required, and any further environmental properties and constraints.

The Basic Platform Modelneeds to lead to an executable Simulation
Platform, and so requires inclusion of initialisation and termination behaviours: the Simulation Experiment Model (152) provides appropriate
specific values for each run.

Simulation Experiment Model

Intent

Define relevant experiments in the simulation, analogous to domain
experiments.

Summary

- build a model to support running Simulation Experiment (177)s that
  are analogues of domain experiments
- design a simulation experiment initialisation approach
- design experiment instrumentation and logging

Context

Part of the Platform Model (149); the computational counterpart of the Domain Experiment Model (135).

Discussion

The Simulation Experiment Model details how the Simulation Platform is to be used to run Simulation Experiments, in a manner that allows the results to be compared to Domain results, through the Results Model (174). It includes interfaces to allow user access to simulation, instrumentation to provide outputs for the Results Model, and possibly visualisation of these outputs through the Visualisation Model (181).

The Simulation Experiment Model includes:

- what the explicit parameters of the system are, and how these might be controlled (for example, fixed constants or variables)
- suitable ranges of operation of the parameters (for example, sensible parameter ranges or agent numbers)
- termination conditions for experiments (for example, time condition);
- approach to simulating randomness, derived from the Stochasticity Model (154), including the relevant distributions

One addition from the analogous Domain Experiment Model needed in the Simulation Experiment Model is handling *initialisation* of the simulation. In a natural biological system, for example, the experimental subject is usually some mature organism or tissue. The simulation may need to start from some "immature" basis and "grow" to a mature system prior to the experiment start, or it may be possible to start it as an already mature system, possibly with the need for Calibration (163). It may be initialised in a random setting, an arbitrary setting, a specific 'initial condition', or a physically-realistic setting. It may be initialised to a single common setting for all runs, or to a range of settings. A run may be considered to start at timepoint zero, or after a sufficient time to allow transient behaviour to subside. Choices need to be justified (possibly as a result of preliminary experimentation) and recorded for argumentation purposes.

The Simulation Experiment Model also captures the necessary *instrumentation*, monitoring the core behavioural concepts that are provided by the Domain Model. Concepts in the Domain Experiment Model get translated into Simulation Experiment Model concepts such as:

- experiment manager, that determines how parameters can change across replicate runs
- instrumentation to collect, measure and process data from simulation experiments; this can include:
  - visualisers and data loggers
  - Simulation Platform variables and outputs to measure and record
  - random number seeds
  - conditions that determine the when and how to record and process certain statistics

Related patterns

Outputs need to interface to the Results Model (174) analyses, and to the Visualisation Model (181)

Stochasticity Model

Intent

Model any required stochasticities explicitly.

Summary

- characterise the relevant distributions, using real world data

Context

Modelling stochasticity in the Domain Model (128) and Platform Model (149).

Discussion

The Domain almost certainly contains stochastic properties, due either to true non-determinism, entropy, uncertainty, or the inability to measure in enough detail. These properties can be modelled as randomness in the Platform Model.

Modelling is needed to capture the relevant distributions of the stochastic properties. Pseudo-random number generators (PRNGs) provided in programming language libraries can produce a variety of distributions. An inappropriate choice of distribution can lead to skewed results. Domain sources of entropy can have various kinds of distribution, including, but not limited to, normal (gaussian), exponential, or power law. Where possible, you should use Domain data to characterise the necessary distribution(s).

Different parts of the model may require different distributions, or different parameter values.

The Simulation Platform (161) should exhibit reproducibility. The model may be stochastic, but the simulation should be deterministic, so that runs can be replayed if necessary. The log of a Simulation Experiment (177) should include the relevant PRNG seeds. If there are multiple processes each needing its own stream of random numbers, a further process may need to be implemented to handle the distribution and recording of suitable seeds. If the Simulation Platform is distributed over a cluster, further care needs to be taken to ensure determinism between processors.

Related patterns

Calibration (163), Sensitivity Analysis (175)

Space Model

Intent

Define how physical space is modelled.

Summary

- decide on suitable model(s) for physical space
- relate spatial Domain Model (128) parameters and values in the Data Dictionary (132) to their Platform Model (149) counterparts
- Document Assumptions (108)

Context

Design decisions when moving from Domain Model (128) to Platform Model (149).

Discussion

Choosing a good model of space is a critical step in a simulation: too fine grained, and the simulation is inefficient; too coarse grained, and it has quantisation artefacts; the wrong topology, and it is impossible to calibrate.

Potential models include:

- aspatial, or well-mixed
- spatial
    - continuous: 1D, 2D, or 3D
    - discrete
        · regular: lattice (grid) topology
        · irregular: general network topology
- hierarchical: for example, a network where each node is a (sub)space; or a tree used to model membrane structures

Aspatial, or **well-mixed**, models are arguably the simplest. If key parameter values are constant across physical space, an aspatial model may be sufficient. For example, ODE models are aspatial: they describe only change through time, not change across space.

However, if properties vary across physical space, this may need to be captured in the simulation. Behaviours are possible in spatial environments that are not possible in aspatial ones: different reactions or interactions can occur at the same time in different places because of their localisation in different environments and states. This includes examples such as a reaction-diffusion system making spatial patterns because of spatially varying concentrations, and different reactions occurring in different compartments separated by some form of membrane.

There are different **spatial** models appropriate in different cases.

Space may be **continuous** (items can exist at any location) or **discrete** (items are constrained to occupying only grid cells or network nodes).

Continuous space or grids may be may be 1, 2, or 3 dimensional (lines, planes, or volumes). The **dimensionality** is often chosen to be 2D, for computational efficiency reasons, and for ease of visualisation. Faithfully capturing an essentially 3D physical process in a 2D model can be non-trivial, as the dimensionalities have different properties. For example, intersections are different: in 3D paths can cross without intersecting. Volumes and areas scale differently. For example, consider a simulation of a 3D structure such as a vacuole (compartment) in a plant cell. If the real world vacuole occupies, say, half the volume of the cell, how big should the 2D surrogate be? Should it be half the area, to preserve "volumes", or should it preserve lengths? The answers will depend on the questions being asked of the Research Context (119) and Simulation Purpose (121). Determining suitable scaling may require Calibration (163); such calibration may not be possible, and a 3D model may be needed to capture the relevant behaviours.

Discrete spaces can be used to quantise an essentially continuous space (for example, by using a grid space modelling a tissue), or used to partition different kinds of environments into different **network** nodes, modelling the transport between regions through the network graph connectivity, for example, in a Multi-domain Simulation (242).

Where a grid is being used to discretise a continuous space, the grid size needs to be chosen carefully. The grid should be as coarse as possible for efficiency, but not so coarse it introduces quantisation artefacts. Quantisation artefacts can also be introduced due to the grid reducing the symmetry of otherwise-isotropic continuous space. Also, the grid should respect any natural physical spatial scales. Determining an appropriate grid size can be part of Sensitivity Analysis (175).

Spaces can be **hierarchical**, particularly in the case of a Multi-scale Simulation (244), potentially with different spatial models on different length scales. For example, a network space may use different subspace models at each node (for example, one node representing an aspatial region of well-mixed reactions linked to another node of agents diffusing in a continuous subspace). Another form of hierarchical space is a tree representing the structure of regions enclosing other regions, such as a set of nested membranes.

Spatial parameters in the Data Dictionary (132) need to be related to corresponding Platform Model (149) parameters dependent on the model of space chosen. The relationship may be calculable, or may be established via Calibration (163). Calibration against real domain data is typically problematic, either because of the way space has been modelled, or, as is often the case in real biological domains, real world data is uncertain (for example, component sizes and relative sizes are often quite approximately known). The uncertainties and mismatches between real world and simulation dimensionality need to be recorded, so that the effects are recognised even if these cannot be analysed and quantified in any meaningful way.

Related patterns

Shortcuts (215) for moving around in space will be dependent on the model chosen.

See also the related Time Model (158), particularly for calibrating speeds and accelerations.

Time Model

Intent

Define how physical time is modelled.

Summary

- decide on suitable model(s) for physical time
- relate temporal Domain Model (128) parameters and values in the Data Dictionary (132) to their Platform Model (149) counterparts
- Document Assumptions (108)

Context

Design decisions when moving from Domain Model (128) to Platform Model (149).

Discussion

Simulating physical time can be difficult.

Computations take processor time that might not be simply related to the corresponding physical times being simulated. A mechanism that needs a lot of computation to simulate might happen relatively quickly in the physical system, whereas a mechanism that is computed more quickly (especially if using Shortcuts (215)) might happen relatively slowly in the physical system. So computation time cannot necessarily be used as a simple proxy for physical time: a more sophisticated model of time may be needed. The issue is made more complicated in distributed simulation running across a range of processors that potentially run at different speeds.

Simple simulations often ignore this issue, conflating processor time and simulated time, and assume that all agents or processes execute in lockstep, in a single simulated discrete timestep, and enforce a form of synchronisation. More sophisticated simulations might use some form of event scheduling, explicitly simulating physical time and calculating when the next relevant occurs, allowing agent behaviours or processes to occur asynchronously.

Different parts of the model might require different models of physical time, particularly in a Multi-domain Simulation (242) or Multi-scale Simulation (244). For example, a cell that moves according to a random walk may execute a fixed number of steps in each "travel" timestep, but collision detection may need to check its relative location many times during a "travel" time step to avoid the cell occupying the same space as another cell at some point in its travel.

When discretising continuous time, the simulation timestep size needs to be chosen carefully. The timestep should be as large as possible for efficiency, but not so large it introduces quantisation artefacts. The chosen timestep should respect any natural physical and experimental timescales: for example, bio-systems are typically measured at specific time intervals (hourly, daily, weekly); engineered systems have characteristic timescales. Fourier transforms of time-series data can be used to find the highest frequency events of relevance (although beware artefacts due to measurement timescales); the Nyquist-Shannon sampling theorem implies that the maximum timestep should correspond to an update frequency of at least twice this event frequency, $\Delta t \leq 1/2f$, in order to be able to resolve such events. The timestep should also respect any spatial discretisation: for example, the physical size corresponding to a grid cell combined with a speed implies a maximum timestep: agents typically should not cover multiple spatial grid cells in a single timestep. Determining an appropriate timestep can be part of Sensitivity Analysis (175).

Temporal parameters in the Data Dictionary (132) need to be related to corresponding Platform Model (149) parameters dependent on the model of time chosen. The relationship may be calculable, or may be established via Calibration (163).

Related patterns

See also the related Space Model (155), particularly for calibrating speeds and accelerations.

6.3 Implementation patterns

> ### Implementation Approach
>
> ### Intent
>
> Choose an appropriate implementation approach and language.
>
> ### Summary
>
> - determine coding language, development environment, and approach
> - determine which existing libraries and generic simulation frameworks might be used
> - determine testing strategy
>
> ### Context
>
> A component of the Simulation Platform (161) pattern. Chosen in concert with the Modelling Approach (111) for the Platform Model (149).
>
> ### Discussion
>
> The Implementation Approach used to develop the Simulation Platform needs to lead to an application that is traceable back to the Domain Model (128), for it to be possible to Argue Instrument Fit For Purpose (186). So the implementation approach should use Seamless Development (214) where possible, but do not let that compromise building a natural Simulation Platform that is readily arguable as fit for purpose. Determine whether use of existing libraries or frameworks[1] will help or hinder such traceability.
>
> The approach must also be feasible within the resources of the project as laid out in the Research Context (119), particularly the skills of the team, and the effort available.
>
> The approach should be flexible enough to allow a range of related Simulation Experiment (177)s and hypotheses to be addressed during the Exploration Phase (97).
>
> The testing strategy should be considered in the context of needing to Argue Instrument Fit For Purpose. It needs to suitable for the criticality as laid out in the Simulation Purpose (121). Test Driven Develop-

ment (TDD) [27] can provide a suite of well-designed tests that can be incorporated into the argumentation. Specific aspects to consider for simulation testing are:

- stochasticity: different runs naturally give different results, so testing may need to include a statistical analysis over multiple runs; this is related to the Stochasticity Model (154)
- concurrency: if the simulator is implemented as a concurrent application, there may be fairness aspects to consider, to ensure that each process is simulating sufficient activity: this may be related to the Time Model (158)

Related patterns

Consider implementing or using an existing Domain Specific Language (228) where appropriate.

If the domain has parts that would benefit from different modelling or implementation approaches, combine them in a Hybrid Model (212).

Beware *One Size Fits All* (142) if using a pre-determined implementation framework.

Simulation Platform

Intent

Develop the executable simulation platform that can be used to run Simulation Experiment (177)s.

Summary

- choose an Implementation Approach (160) for the platform modelling, following the principle of Seamless Development (214) as much as possible
- coding
- testing
- perform Calibration (163)
- Document Assumptions (108) relevant to the simulation platform
- if necessary, Propagate Changes (168)

Context

A component of the Development Phase (96), with an existing Platform Model (149).

Discussion

This is possibly the most straightforward part of a CoSMoS development project. There is a well-defined Platform Model available, which forms the specification for the code to be implemented and tested. The development should follow traditional software engineering best practices. The process of software engineering elaborates (reifies, refines) the abstract design specified in the Platform Model to a specific code target. The exact changes depend on the target code medium and the level of detail in the platform model.

Some components may be changed from Platform Model to Simulation Platform: there may be a need for further simplifications, refinement of data structures, and so on. There will be additions, as design decisions are made, and detail is added to allow implementation. These require additions to the Data Dictionary (132), and documentation of the relationships, for use in Calibration (163).

The development of the code typically requires at least the following conventional and boilerplate additions, of which there are equivalents in most programming languages. Note that, where a programming language is supported by a development environment, such as Eclipse[2], it is possible to create some of these additions automatically.

- choice of development location (folder, directory), and assignment of appropriate paths and settings to permit compilation
- package and module structure
- data import and export
- choice of libraries
- choice of data structures, variables, etc.
- choice of concurrency mechanisms, as well as which behaviours occur in parallel in the program

There are four main occasions for executing the simulation platform:

- testing and debugging during development
- Calibration (163) runs
- performing Sensitivity Analysis (175)
- running a Simulation Experiment (177)

As for the Exploration Phase (97) and Platform Model, implementation gives rise to design decisions and assumptions. During the process of developing the simulation platform, ambiguities, errors or inadequacies may be discovered in the Domain Model (128) or the Platform Model. If so, it is important to Propagate Changes (168) through all the models and arguments to ensure the whole development is consistent.

In order to Argue Instrument Fit For Purpose (186), there needs ongoing activity to Document Assumptions relevant to the simulation platform. The correctness of the platform may need further analysis: Calibration tunes the simulator, but in order to understand the behaviour of the simulator under different experimental conditions, it is likely to need to conduct Sensitivity Analysis on the parameters. This may lead to further development activity to improve the robustness of the simulator, or to Refactor (233) out components that have no effect on simulation outcomes.

Calibration

Intent

Tune the Simulation Platform (161) so that simulation results match the calibration data provided in the Data Dictionary (132).

Summary

- select calibration data (inputs and outputs)
- select validation data (inputs and outputs)
- calibrate instrument

  - determine a translation from domain input data to simulation input data
  - determine a translation from raw simulation output data to results data
  - compare domain outputs to translated simulation outputs
  - iterate until sufficiently similar

- validate that the calibrated result is not overfitted
- if calibration cannot be achieved, revisit the models and assumptions

Context

A component of the Simulation Platform (161).

Discussion

Calibration is a standard part of the manufacture and deployment of any scientific instrument. It often refers to setting the correct zero point and scale. Physical scientific instruments may need to be recalibrated if environmental conditions change (such as temperature causing expansion of parts of the device). Simulation scientific instruments should only need to be calibrated once before use, but do need to be recalibrated if the simulation platform is changed in any way (see *Tweaking Code* (183)).

Calibration is required in order to bring the Simulation Platform to an experimentation-ready state. Domain parameter values need to be translated to appropriate platform values, and values for proxies and surrogates determined. Uncertainties in parameters (and potentially in sub-models) can be addressed by exploring the parameter space (or trying different sub-models). The aim is to obtain outputs from the Simulation Platform that are in agreement with calibration data, without overfitting. Calibration can be performed through simple, manual tuning or more elaborate methods, such as gradient techniques or evolutionary algorithms. Multi-objective optimisation can be used to calibrate to a range of performance metrics [186].

The various kinds of data involved are part of the Data Dictionary (132). Figure 5.3 (p.134) shows the various data components in detail. The calibration data is used to adjust the translations and parameter values until the Results Model (174) data fits the Domain Model (128) experimental results as captured in the Domain Experiment Model (135). The validation data is used to ensure the model has not overfitted the calibration data.

The Domain Model has input data, including model parameter values in the Basic Domain Model (130), and experimental parameter values in the Domain Experiment Model (135). A domain experiment based on this experimental data produces raw output data. After the appropriate scientific data analyses, this yields the domain results data.

To move to the simulated world, the domain data needs to be translated to appropriate Simulation Platform values. A simulation experiment given simulation input data (simulation parameters and experimental setup) produces raw simulation data. This needs to be translated back into domain world terms, and then similarly analysed in the Results Model, to yield the simulation results. The translation and analyses are designed so that the simulation results in the Results Model are directly comparable to the domain results.

The Calibration exercise is to adjust the translations to simulation inputs and from simulation outputs to achieve suitably similar simulation and domain results, on the calibration data. Even if you have values in the Domain Model for most of these parameters, they often come from different sources (for example, animal and human, different laboratories) and from a real world system most of which you have abstracted or outright ignored in the simulation. This means that some adjustment can be expected in aligning simulation results with Domain Behaviours. For some parameters, you may not have values available in the the Domain Model because they are not known; then Calibration becomes not only translation, but finding appropriate values.

The relationship between domain and simulated results need not be exact equality, but can be statistical similarity. The domain and simulation experiments are not functions in the mathematical sense, since different experimental runs on the "same" input data will yield different output data, due to variation, experimental error, and stochasticity. This similarity should be compared by analysing the simulated outputs using the same statistical tests as in the Domain Experiment Model (135), for comparability. For example, Domain experiments typically have control and experimental groups, between which there is a difference in emergent behaviours, detected through some statistical test. Do analogous experiments in simulation, and observe the difference between control and experimental groups through the same statistical test. This does not compare results data directly between the domain and simulation experiments, but compares the results of the same statistical test.

Consideration should also be given to calibrating the initial conditions. A simulation often has an initial state, whereas the domain being simulated is often an ongoing process. Suitable initial states need to be captured and calibrated, or the amount of running time needed to go take the system beyond transient behaviour needs to be determined.

The translation from domain to simulation values may be relatively trivial (not much more than the identity transformation) if the Domain Model and Platform Model are very similar. However, it might be sophisticated, if the Platform Model has introduced differences, such as surrogate entities standing in for multiple Domain entities, change of spatial dimension (see Space Model (155)), non-trivial discretisation, and so on. If information is lost by the translation, it will not be available during analysis of the simulation results, and so should not be critical in the analysis of the domain experiment results either.

Where the Domain Experiment Model (135) results are in terms of emergent properties, the Domain Scientist (101) will need to agree on a suitable translation from simulation micro-states to appropriate macro-states. This process will need to consider the Domain Behaviours (135) model and the Research Context (119), and be captured in the Simulation Behaviours (179) model. This can be challenging when the emergent properties observed in the Domain are many levels and scales away from the micro-state properties being simulated: what levels of simulated chemicals in a simulated tissue correspond to the observed "mouse is sick; mouse died"? Metrics acting as *in silico* proxies for these emergent behaviour observations *themselves* require calibration, and this must be performed on a calibrated simulation[3]. It may be necessary to change some aspects of the Simulation Purpose (121), or identify new domain data, in order to calibrate the system.

One technique that can be used to help calibrate surrogates is to express parameters in terms of dimensionless quantities (for example, the Rayleigh Number or the Reynolds Number) to minimise the effect of unit choices and other changes.

Calibration data should be selected to ensure good calibration coverage of the domain of interest, encompassing the range of possible outcomes, and of relevant experiment types. The system should not be used "out of calibration", that is, in an area of experimental space not well covered by the calibration data; this should form part of the fitness-for-purpose argument. If a system is used out of calibration, for extrapolation, the resulting predictions should be checked by further domain experiments.

Calibration is a "data-fitting" process: translation is tuned so that the simulation adequately reproduces the calibration data. As such, common data-fitting issues such as "overfitting" need to be avoided. In particular, the *form* of the translation should not be arbitrarily fitted; its design should be constrained and guided by the kinds of changes made moving from Domain Model to Platform Model. Validation data, providing the same coverage as the calibration data, should be used to test that the calibration is not over-fitted; this should form part of the fitness-for-purpose argument.

If the model is fit for purpose, Calibration should be achievable by setting the relevant parameter values (for example, of surrogates). However, the calibration exercise may demonstrate that the model is inadequate in some way, for example, a missing behaviour, and it cannot be calibrated to produce suitable output. The response to this is to return to the Discovery Phase (95), revisit and rework the underlying models to discover the holes, possibly with the aid of a Prototype (213), Propagate Changes (168), and recalibrate the new system.

Tweaking Code (183), on the other hand, is "forcing" the model to fit the data, for example, by using domain-implausible parameter values (for example, many orders of magnitude away from real world values), or altering the code, independent of the models, to make it "work". Where the over-arching purpose of the simulation exercise is to understand the Domain science, failure to calibrate is an opportunity to improve this understanding, not a reason to paper over the holes.

Related patterns

Calibration is distinct from Sensitivity Analysis (175), but the latter can be used to help determine how rigorous calibration need be for parameters with ill-defined domain values.

Translation from, say, 3D physical space to a 2D simulated Space Model (155) may need to be performed through Calibration.

Translation of time from domain time to simulated time can be captured in the Time Model (158), and refined during Calibration.

Debug By Video

Intent

Use visualisation to help debug the simulation platform.

Summary

Implement the Visualisation Model (181) as early as possible, and use it to help debug the implementation.

Context

Debugging the implementation of the Simulation Platform (161).

Discussion

Visualisation can be useful for visual debugging [102]. It can provide a rapid intuition of what is happening, and help build confidence that the simulation is not completely wrong. The human visual system is good at spotting subtle problems and patterns that are difficult to quantify, even if you know to look for them. The Domain Scientist (101) should be recruited to spot if things look wrong in Domain terms.

Related patterns

Not to be confused with *Proof by Video* (182).

Propagate Changes

Intent

Ensure that changes in one part of the system propagate throughout, to ensure consistency.

Summary

- identify the original source of the problem, and fix it there
- propagate the fix throughout the system models and arguments
- make a large fix through a series of small independents steps, and propagate small changes only, to keep the process manageable

Context

Whenever development of a particular model or argument requires changes to other models or arguments.

Discussion

If a problem is found during development, for example when implementing the Simulation Platform, there is a great temptation to fix it there and then. However, the entire rationale for the CoSMoS process is to have models and arguments that ensure the simulation is fit for purpose. So these models and arguments must all be describing the same system. An error discovered in one phase or model may well be due to a problem in a different phase or model. It is essential to fix the problem where it originates, and to propagate the changes due to that fix throughout the system.

Don't make radical changes to one part, then try to propagate; instead, make small individual changes throughout the system. This is analogous to the Refactor (233) philosophy of improvements through a controlled series of small changes, and to the idea of incremental development through a controlled series of small additions, as in a Multi-increment Simulation (240).

Related patterns

Use Version Control (221) to manage the different versions of the models, software, and experiments.

Tweaking Code (183) can also lead to problems of models not being in sync.

6.4 Antipatterns

Amateur Coding

Problem

You believe that "anyone can write code".

Context

The Domain Scientist (101) building the Platform Model (149) and Simulation Platform (161).

Discussion

After all, how hard can it be?

If you are finding it difficult to Document Assumptions (108) about the Platform Model or Simulation Platform, you may be engaged in *Amateur Coding*.

Compare *Amateur Science* (137) for the other side of this anti-pattern.

Solution

Realise that in all but the simplest simulations, software engineering expertise is needed.

Premature Implementation

Problem

You start writing Simulation Platform (161) code before having a proper understanding of the Domain or platform requirements.

Context

During Discovery Phase (95) or Development Phase (96).

Discussion

Also known as "hacking". Such leaping into code before the requirements (Domain Model (128)) or simulation architecture (Platform Model (149)) are properly understood leads to opaque simulators that are impossible to argue fit-for-purpose, and to the inability to build or interpret the Results Model (174) in any meaningful manner.

Solution

Remember to Argue Instrument Fit For Purpose (186): you will not be able to if there is a significant code with no corresponding platform model. If you really need to explore coding or other ideas, Prototype (213).

Beware of the opposite anti-pattern: *Analysis Paralysis* (138). A related antipattern is *Tweaking Code* (183).

Program In the Answer

Problem

The results from the Simulation Experiment (177) are an inevitable consequence of the simulation programming, not an emergent consequence of the operation of the simulation.

Context

Building the Platform Model (149) and Simulation Platform (161).

Discussion

The simulator is being used to explore scientific questions, such as "does this hypothesised set of behaviours lead to that observed set of outcomes"? If the observed set of outcomes are themselves programmed in to the simulator, independent of the hypothesised behaviours, then a misleading answer will be inferred.

Solution

Have a careful separation of concerns in the Domain Model (128) to identify what the required answers are, captured in the Domain Behaviours (135), and ensure that they are not available to the Platform Model (149) or to the simulation.

Independent Simulation Implementor

Problem

The Simulation Engineer (105) diverges from the Domain Model (128) in order to create a more aesthetic or efficient computer program, and produces a program that breaks the Seamless Development (214) on which the engineering fitness for purpose relies.

Context

Argue Instrument Fit For Purpose (186): reviewing arguments to ensure consensus

Discussion

Computer system developers are often expert programmers, who know many idioms or styles that can create efficient computational solutions that produce the required outputs. However, when simulating a complex system, it is usually important that the mechanism creating the outputs bears some relationship to the mechanism in the simulated system. It is hard to complete a fitness-for-purpose argument if it cannot be shown how the structures and behaviours of the simulation map the structures and behaviours of the subject of simulation.

Note: there are some short-term prediction simulations where the fitness-for-purpose depends only on mirroring the output patterns of the simulated system. In this case, part of the purpose may relate to the speed of computation, and the Seamless Development (214) is less important than the computer performance. In such a case, the fitness-for-purpose argument needs to reveal the limitations of the results, particularly the likelihood that simulation and reality may significantly diverge beyond the very-short-term.

Solution

If the Simulation Engineer believes that there is an efficient implementation which is inconsistent with the Domain Model, it may be possible to revisit the Domain Model, making changes that are consistent with the efficient implementation. However, any changes must be consistent with the Domain, and must be revisited with the Domain Modeller and the Domain Scientist: the Structured Argument that the appropriate instrument has been designed must be re-evaluated.

If it is not possible to accommodate the efficient implementation, it is essential that the quality of software engineering (for example, Seamless Development) takes precedence over the performance of the computer program. A slow program that gives usable results will always be better than a fast program that gives untrustworthy results.

Chapter 7
Exploration phase patterns: using the platform

Abstract — In which we describe: using the simulation platform as a scientific instrument appropriately; running simulation experiments; patterns for the results model.

7.1 Catalogue of exploration phase patterns

Model and usage patterns	
Results Model (174)	Build an explicit description of the use of, and observations from, the Simulation Platform.
Sensitivity Analysis (175)	Determine how sensitively the simulation output values depend on the input and modelling parameter values.
Simulation Experiment (177)	Design, run, and analyse simulation experiments.
Simulation Behaviours (179)	Develop a model of the emergent properties of a simulation experiment, for comparison with the related Domain Behaviours model.
Visualisation Model (181)	Visualise the simulation experiment results in a manner relevant to the users.

© Springer Nature Switzerland AG 2018
S. Stepney, F.A.C. Polack, *Engineering Simulations as Scientific Instruments: A Pattern Language*, https://doi.org/10.1007/978-3-030-01938-9_7

Antipatterns

Proof by Video (182)	The visualisation model is all there is.
Tweaking Code (183)	You make a series of small, "unimportant" changes to the working Simulation Platform.
Tweaking Experiments (184)	You make a series of small, "unimportant" changes to the defined Simulation Experiment.

7.2 Model and usage patterns

Results Model

Intent

Build an explicit description of the use of, and observations from, the Simulation Platform (161).

Summary

- perform Sensitivity Analysis (175)
- perform relevant Simulation Experiment (177)s
- build a Simulation Behaviours (179) model

Context

A component of the Development Phase (96), with an existing Simulation Platform (161).

Discussion

The Results Model is a descriptive model of the *simulation* domain, as understood from Simulation Experiments and observations. It contextualises the simulator output in a way that makes it useful and informative to the Domain Scientist (101), but it must avoid making the simulation results seem more reliable or authoritative than they are.

It is built from Simulation Experiment data in an analogous manner to how the Domain Model (128) is built from the Domain (123) data, and focusses on the emergent properties (Figure 1.3, p.21). Sensitivity Analysis (175) also provides input data to this process.

Sensitivity Analysis

Intent

Determine how sensitively the simulation output values depend on the input and modelling parameter values.

Summary

Clearly identify the purpose of the analysis: there are many SA techniques, each of them capable of revealing different information. Consult the technical literature to perform the appropriate technique.

Context

Arguing the Results Model (174); Calibration (163).

Discussion

Sensitivity Analysis is an umbrella term given to statistical methods that discover how perturbation of a system's inputs is reflected in its outputs. Sensitivity Analysis can be used to quantify how influential particular simulation parameters are on simulation output, and what range of output values results from a range of input values[4].

Whilst Sensitivity Analysis operates over a simulation's parameters, its results may be considered representative of how influential simulation components are. Parameters provide values for the rates, probabilities, quantities and timings concerning components and their interactions. If a parameter is found to be influential, this indicates that the respective component and/or interaction is influential.

Sensitivity Analysis has a variety of uses in complex simulation.

- It has a role in checking the Domain Model (128). If it is known which components of the domain are highly influential (or otherwise), similar patterns should be reflected in the simulation. Discrepancies between influences of corresponding real world and simulation components motivates further investigation to ensure that the Domain Model, Platform Model (149), and Simulation Platform (161) are in fact representative of the real world Domain (123).
- Sensitivity Analysis results can inform Calibration (163), highlighting influential parameters and the direction of correlation between parameter adjustments and the effect on simulation behaviours.

Where a simulation requires adjustment to better align its behaviour with the Domain, Sensitivity Analysis can reveal which parameters to adjust, and in which direction.

- Results of Sensitivity Analysis form an important part of the Results Model. Along with an understanding of the uncertainty concerning particular aspects of the Domain, and hence simulation parameters, Sensitivity Analysis results that highlight the influence of parameters on simulation behaviour indicate how representative those behaviours are of the Domain. If highly influential simulation parameters cannot be specified because of uncertainty in the Domain, then simulation results may simply represent parameterisation artefacts. These results are important when attributing confidence to simulation results, and are an important component of the related argument.

- Sensitivity Analysis represents a powerful means for exploring a simulation's dynamics, and generating hypotheses from this. Applied at various times during simulation execution, Sensitivity Analysis can highlight how the influence of particular simulation components changes over time.

- Results of Sensitivity Analysis may also help in understanding how the model could be simplified. Where the analyses suggest components that have no effect on simulation behaviour when the parameter value is perturbed, this may direct statistical analyses to focus in other areas, saving execution time, and provide reasoning to the simplification of that component.

The application of Sensitivity Analysis requires that simulation responses be defined: the specific outputs of the simulation that are analysed whilst perturbing its inputs. Any aspect of simulation behaviour can be selected as a response, there is no limit on the number of responses that can be selected, and as demonstrated in the examples of use below, responses can be elaborate metrics that require calibration.

Sensitivity analyses may be broadly categorised as being *local* (one-at-a-time, or OAT) analyses or *global* analyses. Local techniques vary and analyse a single parameter at a time, holding all other parameters at default values. Global techniques simultaneously perturb multiple parameters during analysis. Since local techniques perturb only a single parameter at a time, only first-order sensitivities are revealed. Global techniques can reveal sensitivities where one parameter's influence depends on the value held by another. This does not render global techniques necessarily superior: the choice in technique should be driven by the information that is to be acquired. In a stochastic simulation in particular, local techniques can provide more accurate information relating to particular aspects of a single parameter's influence. When applied to stochastic simulations, global techniques are subject to considerable variation in the data, since the inherent stochasticity in the simulation is compounded with the fact that multiple influential parameters are randomly perturbed by substantial quantities.

Multiple runs, requiring considerable computational power, may be required to ensure that patterns observed through Sensitivity Analysis are not merely due to stochasticity. Such multiple runs provide the opportunity to analyse variance in simulation behaviours. Variance may not be equal at all points in parameter space.

Sensitivity Analysis can be qualitative as well as quantitative. For example, HAZOP (Hazard and Operability) and similar studies perturb systems qualitatively, with probes such as 'negation', 'less', 'more', 'part', 'reverse'. These can be viewed as a qualitative Sensitivity Analysis technique, and can be adapted to certain aspects of simulation.

Saltelli et al. [199] provide a comprehensive review of sensitivity analysis techniques. An example of how using Sensitivity Analysis within a CoSMoS Simulation Project resulted in a novel scientific discovery is provided in the endnotes[5]. We have developed an open source R implementation that performs many of the analyses discussed here [4, 6].

Related patterns

Contrast with Calibration (163).

Document Assumptions (108): an assumption might rely on Sensitivity Analysis for the justification.

Sensitivity Analysis can be used to design appropriate grid sizes in the Space Model (155) and Time Model (158).

Simulation Experiment

Intent

Design, run, and analyse simulation experiments.

Summary

- design the experiment
- perform simulation runs and gather data
- analyse results, for input to the Simulation Behaviours (179) model
- Document Assumptions (108) relevant to the simulation experiment

Context

A component of the Results Model (174). Using the Simulation Platform (161) in the context of a Simulation Experiment Model (152).

Discussion

Using the Simulation Platform to run Simulation Experiments should be approached in the same way as running experiments in the Domain: the experiments need to be designed, run, and the results analysed in an analogous manner.

Detailed advice on the conduct of experiments is outside the scope of this work. Certain aspects are CoSMoS-specific, however. The Simulation Experiment Model provides the basis for the design and running, and the Simulation Behaviours (179) model provides the basis for the analysis.

The Simulation Experiment Model provides a model for simulation experiments that are appropriate analogues of Domain experiments. For each individual experiment, we need to consider the specific instantiation of the model. This includes:

Design: Instantiating the Simulation Experiment Model as a single Simulation Experiment instance:

Initialisation: The initial state of the simulation run.

Transient behaviour: When data collection starts, possibly after some initial transient behaviour that should be ignored while the system "settles down".

Number of simulation runs: Different authors have various philosophies and give various advice [43, 193]. A traditional reasoning would be based on the required statistical significance, statistical power, and effect size. Argumentation should include consideration of the number of domain experiment replicates. Calibration (163) runs can be used to determine distributions, and whether they satisfy the normality requirements of parametric statistics.

Run: Executing the Simulation Experiment:

Parameters: The specific parameter values for this run.

Repeatability: Repeatability is crucial. The Stochasticity Model (154) covers where randomness is included, and with what distribution. Run records should record the relevant random seeds used, to ensure repeatability. The Simulation Experiment includes experiment logging requirements to ensure repeatability.

Termination: When the run is considered complete, and data collection stops.

Analyse: Analysing the results of the Simulation Experiment:

Calibration (163) defines how the raw simulation outputs can be transformed to data comparable with domain output data. The simulation outputs should be analysed using the same techniques as the domain data, to allow comparison. The Simulation Behaviours model provides the requirements for this.

Simulation Behaviours

Intent

Develop a model of the emergent properties of a Simulation Experiment (177), for comparison with the related emergent Domain Behaviours (135) of the Domain Model (128).

Summary

- build a *minimal* model, from consideration of the Research Context (119), the Simulation Experiment Model (152), the Domain Behaviours (135), and the Calibration (163) translation of the raw simulation data
- if needed, build an *augmented* model including micro-level observations, and argue the connection to the domain model data
- if needed, build a Visualisation Model (181)

Context

A component of the Results Model (174).

Discussion

This model is the simulation analogue of the domain model's Domain Behaviours.

The Calibration (163) translation (Figure 5.3, p.134) says how to translate simulation raw output data into a form that can be compared with domain data. The Simulation Behaviours model is built using this transformation, along with the analyses captured in the Simulation Experiment Model (152), to present Simulation Experiment results in form that is comparable to Domain experiment results. The form of the comparison is dictated by the Research Context (119) and Simulation Purpose (121): what hypotheses is the Simulation Experiment testing?

The comparison requires that the analysis be performed using the same statistical tests as used for the Domain. However, there are deep issues with standard statistical testing, from the use of parametric statistics on inappropriate distributions, to the whole concept of statistical significance [147], and the need to calculate effect sizes. So do not blindly follow what the Domain Scientist (101) uses: challenge it, and add a justification to the arguments. Where there are deep disagreements over the correct test to use, one compromise is simply to document the results of both tests, for example, a parametric and a nonparametric one.

A *minimal* Simulation Behaviours model maps directly on to the Domain Behaviours (135), or emergent properties, model (Figure 1.3, p.21). The Results Model does not contain an explicit analogue of the Basic Domain Model (130); there is no explicit model of the simulated agent-level behaviours.

However, simulation allows access to much more data, particularly of micro-level behaviours and values. An *augmented* Simulation Behaviours model can contains such extra information, which may allow more analysis and conclusions compared to what is possible with Domain experiments. An explicit relationship must be defined between the Research Context and such an extended results model, to give principled guidance for the analysis. This relationship will need to be argued.

It may not be possible to make a direct mapping between the emergent Simulation Behaviours and the emergent Domain Behaviours (135), because these may be at different emergent levels. For example, the simulated emergent behaviours may be at the level of a tissue (with the agents at the level of cells), whilst the domain emergent behaviours may be observed at the level of the whole organism (at the most extreme, whether it dies or not). In such cases, it will be necessary to develop further analysis approaches to translate the observed Simulation Behaviours into results that are meaningful in the Domain [185, 188]. This will also give rise to separate arguments of appropriateness and fitness for purpose.

Related patterns

The simulation data may be presented in a visual form too, via a Visualisation Model (181).

Visualisation Model

Intent

Visualise the Simulation Experiment (177) results in a manner relevant to the users.

Summary

Model how the experimental data from the Domain (123) is presented to the user. Present the simulation results in a similar way in the Results Model (174).

Context

A component of the Simulation Behaviours (179).

Discussion

Domain experiment results can be presented in a wide range of forms, such as tables, graphs, charts, maps, images, animations. Use an analogous style when presenting the Simulation Experiment results (in addition to providing the actual data), to help the Domain Scientist (101) interpret what the simulation is showing. Proper analysis is also needed: just because the results superficially look the same does not necessarily mean that they are the same.

Any such visualisation should be scientifically sound, to build visualisations that communicate the correct information effectively. One approach is using techniques from *visual analytics* [135]. Interactive visualisation can provide potential added value in the ability to explore the simulation data at depth.

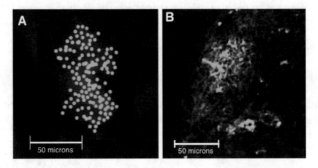

Fig. 7.1 A visualisation model example: A) simulation visualisation of a Peyer's Patch; B) actual confocal microscopy image. Image from [41, Fig. 3]

Related patterns

Visualisation can be used to Debug by Video (167). But beware of *Proof by Video* (182).

7.3 Antipatterns

Proof By Video

Problem

The Visualisation Model (181) is all there is.

Context

Using a Visualisation Model (181) as part of the Results Model (174).

Discussion

The visualised results from the simulation look superficially similar to the visualised results from the Domain (be it a static figure or an animation), and so you judge the simulation to be a "success". But there is no quantification of the similarity of the results, so you cannot be sure the correspondence is more than an optical illusion, and you cannot make any quantitative statements or predictions.

Solution

Define an appropriate Calibration (163) and Results Model (174), and rigorously compare simulation output with Domain experimental values.
Not to be confused with Debug by Video (167).

Tweaking Code

Problem

You make a series of small, "unimportant" changes to the working Simulation Platform (161).

Context

Running a Simulation Experiment (177).

Discussion

You have a working, calibrated, validated Simulation Platform (161), and are performing Simulation Experiments. You notice a small problem, or a place where an improvement could be made, and "tweak" the code. Your Simulation Platform is now slightly out of sych with the models and arguments. The changes are small enough that you do not think that this is a problem. Then you notice an opportunity another small improvement. And another. Very soon the Simulation Platform has drifted far enough away from the models and arguments that you are no longer justified in having confidence in the results, but since each change has been small, you do not realise this.

In actuality, *any* change to the Simulation Platform needs to be retested, recalibrated, and revalidated. This can need anything from a trivial walk-through, to a full scale Calibration (163) exercise, depending on how many assumptions are changed.

Experimental results from different versions of the Simulation Platform should *never* be combined, as there may be small but systematic differences in the results.

Solution

Refactor (233) the Simulation Platform to make disciplined improvements that do not change its functionality. This minimises the need for revalidation and recalibration.

Propagate Changes (168) through the system.

Use Version Control (221) to associate experimental parameters and results with the particular Simulation Platform version.

Combine experimental results only if they are from the same Simulation Platform version. Rerun experiments as necessary.

A related antipattern is *Premature Implementation* (170).

Tweaking Experiments

Problem

You make a series of small, "unimportant" changes to the defined Simulation Experiment (177).

Context

Running a Simulation Experiment (177).

Discussion

You can run your simulator to get a lot of results. You find it easy to do some tweaking of the experimental parameters, to get some more data. You obtain many results for which you do not have associated Domain data. You find yourself eventually with high volume, low confidence data, not easy to publish or analyse.

Solution

Be methodical. Ensure that each simulation run is part of a well-defined Simulation Experiment, with well-defined goals.

Chapter 8
Structured argumentation patterns

Abstract — In which we describe patterns for developing structured arguments using a specific notation.

8.1 Catalogue of argumentation patterns

Argument patterns	
Argue Instrument Fit For Purpose (186)	Provide an argument that the CoSMoS Simulation Project is fit for purpose.
Structured Argument (189)	Structure and develop the required arguments in a systematic manner.

Basic Structured Argument patterns	
Argument Structuring Notation (190)	Provide a diagrammatic summary of an argument, to highlight the structure, and provide an index into the detailed argument.
Claim (192)	Represent a claim in an argument.
Strategy (194)	Outline how a Claim will be substantiated; the reasoning step that shows that the sub-claims would substantiate a claim.
Evidence (195)	Indicate evidence that substantiates a Claim.

© Springer Nature Switzerland AG 2018
S. Stepney, F.A.C. Polack, *Engineering Simulations as Scientific Instruments: A Pattern Language*, https://doi.org/10.1007/978-3-030-01938-9_8

Argument Context (196)	Provide information about the context in which a Claim or Strategy should be interpreted.
Assumption (197)	Record assumptions that must hold for a Claim or Strategy.
Justification (198)	Record the justification for a Claim or Strategy.
Unfit (199)	Record a development weakness in an argument step.

Generic Argument patterns

Create Generic Argument (200)	Generalise part of a Structured Argument, so that it can be instantiated in different contexts.
Use Generic Argument (202)	Instantiate a generic argument for use in a specific context.

8.2 Argument patterns

A computer simulation is developed for a purpose. It is important to consider its fitness for its intended purpose. We use structured arguments to capture the consensus of Domain Scientist (101), Domain Modeller (103), and Simulation Engineer (105) that the simulation is fit for purpose.

In this chapter, we present patterns for one particular approach to presentation of structured arguments, which has been adapted from critical systems engineering and, particularly, safety case argumentation.

Argue Instrument Fit For Purpose

Intent

Provide an argument that the CoSMoS Simulation Project (92) is fit for purpose.

Summary

- establish the fitness-for-purpose *claim*, from the intended purpose of the simulation, as recorded in the Research Context (119)

- establish the required *rigour* of the argument, as recorded in the Simulation Purpose (121)
- agree a *strategy* for substantiating the fitness-for-purpose claim
- use a Structured Argument (189) to substantiate the fitness-for-purpose claim

Context

Part of the overall CoSMoS Simulation Project (92). The responsibility of the Argument Modeller (106).

Discussion

The aim in creating an argument is to express the rationale for the development and use of the simulation (e.g. Domain Scientist (101), Domain Modeller (103), and Simulation Engineer (105)). The argument may be used simply as an internal reference for assumptions, what was decided etc., or it may be exposed to external scrutiny. The argument may also form a review mechanism, to expose limitations or areas where improved confidence is needed.

An important aspect of a simulation fitness-for-purpose argument is that, if the Simulation Purpose (121) changes, the argument must be revisited. The purpose must therefore capture the context of the simulation as well as its intended use: a critical simulation that will be used as primary evidence in research needs a much stronger fitness-for-purpose argument than a simulation that is used to generate or explore hypotheses that can be confirmed by other means (e.g. by conventional laboratory experimentation). Similarly, if the results of a simulation are likely to have significant external impact, then a much stronger fitness-for-purpose argument is needed than if the simulation results are only of interest "internally", to the Domain Scientist.

An argument may be developed **incrementally**, alongside the development of the simulation, or may be developed **retrospectively**, after development (complete or partial). An argument may act as an analysis guide for the development, highlighting areas of concern, the need to explore the effect of assumptions, or areas where some of those involved are not convinced of fitness for purpose.

A fitness-for-purpose argument developed incrementally can be a living argument, and can help to direct the ways in which the simulation development proceeds. The advantage of constructing the fitness-for-purpose argument in parallel with development is that all those involved in development (e.g. Domain Scientist, Domain Modeller, and Simulation Engineer) are prompted to Ask [Silly] Questions (217), and to follow approaches that improve their understanding of the Simulation Purpose, Argument Context (196), Assumption (197), and Justification (198).

An argument may be constructed retrospectively in order to analyse an existing simulation possibly not developed using the CoSMoS approach. Such a process will almost certainly reveal holes in the holes in the rationale for some aspect of the simulation design, implementation or use.

Sometimes, a simulation development is itself an exploration of the Domain: the purpose evolves as the simulation develops, because only the detailed exploration involved can identify the right questions to ask through simulation (see for example [187]). In this case, it may be more appropriate to argue fitness-for-purpose retrospectively, either after completion of the Development Phase, or as part of the Exploration Phase.

When we argue for fitness-for-purpose in parallel with development, the argument is constructed with increments in each of the Discovery Phase (95), Development Phase (96), and Exploration Phase (97). Each increment may revisit and revise earlier argument structures. It may be more practical to start the argument later in the discovery phase, and to argue retrospectively up to that point in the development, rather than to argue incrementally from the very start, because the purpose and scope of the simulation may be revised many times in the early stages of discovery.

Related patterns

When building various models, there are patterns whose content forms much of the rationale for arguing claims, including Research Context (119), Simulation Purpose (121), Document Assumptions (108), choice of Modelling Approach (111), choice of Implementation Approach (160), Calibration (163), Sensitivity Analysis (175), and Propagate Changes (168).

Seamless Development (214) helps to make some claims more readily establishable.

The ODD Protocol (222) provides much content to help establish implementation claims.

Structured Argument

Intent

Structure and develop the required arguments in a systematic manner.

Summary

- use the argument strategy agreed from Argue Instrument Fit For Purpose (186) to establish sub-claims.
- for any sub-claim that can be immediately substantiated with evidence, indicate a *resolved* claim, with reference the appropriate evidence
- for a sub-claim that cannot be immediately substantiated with evidence, agree a *strategy* and sub-claims for resolving it
- iterate the process of establishing strategies, sub-claims, context, assumptions, justifications etc. until all claims have been resolved, or have been agreed to be left unresolved
- use an agreed notation to record the structured argument, for example, Argument Structuring Notation (190)

Context

To Argue Instrument Fit For Purpose (186), when presenting an argument, or argument fragment. The responsibility of the Argument Modeller (106).

Discussion

A Structured Argument is an informational structure composed of items of reasoning and evidence, organised in a structured textual or diagrammatic way.

The use of argument for simulation fitness-for-purpose tends to need less rigour than, say, critical system safety case argumentation. Except for the most critical of simulations (established in the Simulation Purpose (121)), it is not necessary to substantiate every claim with Evidence.

A structured argument starts from a top-level Claim that needs to be substantiated. The argument structure lays out the substantiation of the claim as a connected hierarchy of sub-claims representing a chain of reasoning. Structured Arguments make use of the structures of Claim

(192), Strategy (194), and Evidence (195); these can be further detailed through Argument Context (196), Justification (198), and Assumption (197). A Claim may be *resolved* by Evidence, or may decomposed into sub-claims until evidence can directly resolve them, or until an agreement is reached (by the Domain Scientist, Domain Modeller, and Simulation Engineer) that the remaining claims can be left unresolved [176, 180].

Avoid over-rigorous arguments that are unnecessary for the level of criticality. And avoid trivial arguments that do not probe the design or add any other value.

Related patterns

Argument Structuring Notation (190) provides a notation for presenting a structured argument.

8.3 Basic structured argument patterns

Argument Structuring Notation

Intent

Provide a diagrammatic or textual summary of an argument, to highlight the structure, and provide an index into the detailed argument.

Summary

- use the tree-based notation to record the logical structure of a Structured Argument (189)
- use this tree to provide an index into the full Structured Argument (189)

Context

Presenting a full or partial fitness-for-purpose Structured Argument (189).

Discussion

CoSMoS provides an Argument Structuring Notation developed from Goal Structuring Notation (GSN) [107, 136, 230], itself developed to represent safety arguments in the domain of safety-critical systems engineering.

The nature of scientific research is somewhat different from the engineering of safety-critical systems, so we find that a 'goal'-oriented notation is not particularly suited. This is not because research does not have goals, but because the scientific discourse is usually based on *claims*. Safety-critical systems are constructed for the purpose (goal) of being used, and their evaluation ends through a binary decision (to be or not to be given the go-ahead for mass-production), whereas scientific research is open-ended. As Nicolesc says, "Knowledge is forever open" [167]. Goals associate more naturally with "facts", e.g. the system is safe within a given context, rather than with the uncertainties of research. Consequently, we consider an argument as aiming to *resolve a claim* [3, 176, 180, 217].

This discussion does not preclude the use of CoSMoS for building a simulation of an Engineered Domain (238), but there the desired claim and the details of the relevant argument structure will need to be carefully developed.

The key components of the notation are:

- the Claim (192) of some aspect of fitness for purpose
- the Strategy (194) to be used to substantiate the claim
- the Evidence (195) that resolves the claim

The hierarchical claim structure is shown as a tree in which a Claim links to a sub-claim or a Strategy. If a sub-claim is fully resolved, it is linked to the Evidence that resolves it, which terminates that branch of the hierarchy.

In addition, any component can be qualified or annotated with any of the following:

- Argument Context (196)
- Justification (198)
- Assumption (197)

We also adopt GSN conventions for argument modules and generic arguments as appropriate [107].

A tree form (diagrammatic, or textual) is used to show a summary of the structure of the argument, and to act as an index into the detailed body of the argument, cross referenced by the provided identifiers. An example application of this diagrammatic approach can be seen in arguments constructed to support the implementation of a model of immune system development [3].

Related patterns

This notation is an example of a Domain Specific Language (228).

Claim

Intent

Represent a claim in an argument.

Summary

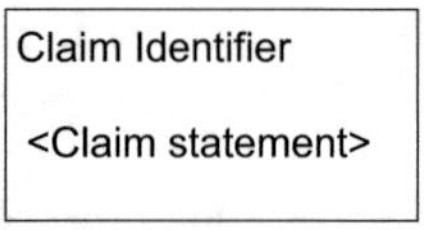

Fig. 8.1 Diagrammatic notation for a claim (based on GSN Goal notation)

Claim Identifier

<Claim statement>

Textual notation for a claim:
[**claim** <Claim id>] <Claim statement>

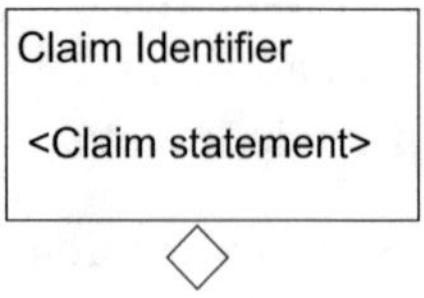

Fig. 8.2 Diagrammatic notation for a claim that is not (yet) expanded further in the argument structure (based on GSN Undeveloped Entity notation)

Claim Identifier

<Claim statement>

◇

Textual notation for an unexpanded claim:
[**claim** <Claim id>] <Claim statement> ◇

Fig. 8.3 Diagrammatic notation for a claim when the argument substantiating that claim is presented elsewhere, as a separate structure

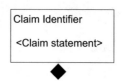

Textual notation for a claim continued elsewhere:
[**claim** <Claim id>] <Claim statement> ◆

Context

In an Structured Argument (189) captured in Argument Structuring Notation (190).

Discussion

The Structured Argument always starts from a Claim. A Claim may have associated Argument Context, Justification and Assumption. The claim can be substantiated by a Strategy or by one or more sub-claims. The notation and semantics of a sub-claim is the same as that for a claim: it forms the start of a sub-argument. Additionally, however, a sub-claim may be associated with the Evidence that substantiates it.

The claim concept is very similar to the goal concept in GSN. Like a goal, a claim may be *incomplete*, and this is represented by a white diamond below the claim. By the end of the argument development, there should be no incomplete claims remaining. However, in our Argument Structuring Notation there is no obligation to fully evidence claims in order to finish the argument structure: it may be the case that those involved in the development (Domain Scientist (101), Domain Modeller (103), and Simulation Engineer (105)) did not consider it important to record further substantiation of this part of the argument; such a claim should be associated with 'Evidence' to this effect.

A Structured Argument may be developed in stages, with incomplete claims being revised, or completed claims revisited. Where a Structured Argument is considered to have become too large, it may be broken down such that incomplete sub-claims in one argument tree become root claims in other trees. The continuation of an argument is indicated in the parent tree by a black diamond attached to the relevant sub-claim.

Strategy

Intent

Outline how a Claim (192) will be substantiated; the reasoning step that shows that the sub-claims would substantiate a claim.

Summary

Fig. 8.4 Diagrammatic notation for the strategy in support of a Claim (based on GSN Strategy notation): the strategy supports the claim, and provides a rationale for the sub-claims

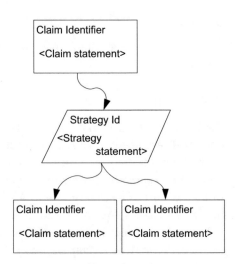

Textual notation for the strategy in support of a claim:
[**claim** <id>] <Claim statement>
   [**strategy** <id>] <Strategy statement>
     [**claim** <id>] <Claim statement>
     [**claim** <id>] <Claim statement>

Context

In an Structured Argument (189) captured in Argument Structuring Notation (190): linked to a Claim (192).

Discussion

Whilst in principle a Claim can be directly broken down into sub-claims, it is useful to record the Strategy for this breakdown. As in GSN,

Strategy is always in support of one Claim, but may be supported by one or more sub-claims. The Strategy should be acceptable to the relevant parties.

The argument fragments associated with particular strategies can form generic argument patterns. For example, an argument of fitness for purpose may be addressed by a three-part strategy over the adequacy of domain capture, the adequacy of the software engineering, and the quality of the simulation results. The three sub-claims that support the fitness for purpose claim are thus rationalised by the strategy.

The Strategy statement is typically worded in the form "Argument over ...", or "Argument by appeal to ...".

Evidence

Intent

Indicate evidence that substantiates a sub-claim.

Summary

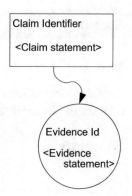

Fig. 8.5 Diagrammatic notation for the evidence that substantiates a claim (based on GSN Evidence notation)

Textual notation for the evidence that substantiates a claim:
[**claim** <id>] <Claim statement>
   [**evidence** <id>] <Evidence statement>

Context

In a Structured Argument (189) captured in Argument Structuring Notation (190): linked to a Claim (192) used as a sub-claim in an argument.

Discussion

An argument is fully substantiated when all its branches end in Evidence: this equates to a GSN argument in which solutions present evidence of the truth of every goal in the structure. In an argument of fitness for purpose of a scientific simulation, it is relatively rare to be able to provide definitive evidence to substantiate a goal, because the science behind many complex systems is incomplete, and the mappings to computational structures are usually "unproven". However, in an engineering context, a simulation may be more amenable to complete fitness for purpose argumentation, with evidence to support each claim.

Evidence represents the substantiation of a Claim in the sense that those involved in the simulation accept this as the end-point. Review of an argument structure may lead to further analysis of the basis of evidence, and extension of the argument: in this case, the original Evidence is replaced by strategy and/or sub-claims.

Argument Context

Intent

Provide information about the context in which a Claim (192) or Strategy (194) should be interpreted.

Summary

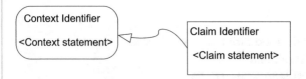

Fig. 8.6 Diagrammatic notation for the context of a claim (or strategy): the context references definitions and other contextual information that is relevant to the claim (or strategy). (Based on GSN Context notation.)

Textual notation for the context of a claim (or strategy):
[**claim** <id>] <Claim statement>
　　[**context** <id>] <Context statement>

Context

In a Structured Argument (189) captured in Argument Structuring Notation (190): linked to a Claim (192) or Strategy (194).

Discussion

A Claim or Strategy may make general statements that need to be explained or elaborated by the Argument Context. For example: a Claim that applies across a range of components may have an Argument Context that states that all the relevant (types of) component(s) have been enumerated; a Strategy that applies across a range of components may have an Argument Context noting that each component must be considered explicitly for the Claim to be substantiated under this strategy.

The normal use of Argument Context in our Argument Structuring Notation follows the GSN syntax, in which a context is only associated with goals and strategies, and is a terminal concept (there cannot be another context, a justification or assumption directly linked to a context).

Assumption

Intent

Record assumptions that must hold for a Claim (192) or Strategy (194).

Summary

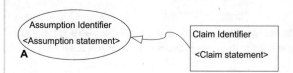

Fig. 8.7 Diagrammatic notation for an assumption associated with a claim (or strategy), used in support of the argument. (Based on GSN Assumption notation.)

Textual notation for an assumption associated with a claim (or strategy):
[**claim** <id>] <Claim statement>
  [**assumption** <id>] <Assumption statement>

Context

In a Structured Argument (189) captured in Argument Structuring Notation (190): linked to a Claim (192) or Strategy (194).

Discussion

A Claim or Strategy may depend on some Assumption to hold. In GSN, a common argument strategy is "Argument by appeal to elimination of all hazards", which makes the Assumption that "All credible hazards have been identified" [107]. Such an assumption may itself be subject to a separate argument. In our Argument Structuring Notation, an argument Assumption may be used in this strict way, or may be simply be an annotation that points out to some documented assumptions (see Document Assumptions (108)) that affect the credibility of the Claim of fitness-for-purpose at this point in the argument.

Justification

Intent

Record the justification for a Claim (192) or Strategy (194).

Summary

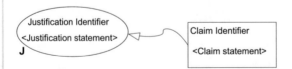

Fig. 8.8 Diagrammatic notation for the justification of a claim (or strategy), which can be used in support of the argument. (Based on GSN Justification notation.)

Textual notation for the justification of a claim (or strategy):
[**claim** <id>] <Claim statement>
  [**justification** <id>] <Justification statement>

Context

In an Structured Argument (189) captured in Argument Structuring Notation (190): linked to a Claim (192) or Strategy (194).

Discussion

A Claim or Strategy may be used in a Structured Argument because those involved in the simulation have some prior knowledge or are following some defined approach. The Justification annotation allows the reason for a Claim or Strategy to be referenced explicitly.

Unfit

Intent

Record a development weakness in an argument step.

Summary

Tag an argument step as indicating the instrument is **!! Unfit**, and explain the weakness.

Context

Any step of a Structured Argument (189).

Discussion

When developing a fitness for purpose argument, it is quite possible that you will discover places where the development has not been sufficiently rigorous to allow the argument to progress.

In the argument, tag the uncovered weakness with **!! Unfit**, explain what the development is lacking, and what needs to be provided to allow the argument to progress. See, for example, the running example on p.88.

8.4 Generic argument patterns

Create Generic Argument

Intent

Generalise part of a Structured Argument (189), so that it can be instantiated in different contexts.

Summary

Following the GSN conventions (used but not defined in [107]), create a generic argument by replacing all specific terms in all relevant components of the argument with appropriate generic terms or parameters.

Context

Argument Structuring Notation (190).

Discussion

It is often the case that an argument structure is generic to particular situations. In this case, it is useful to have a structural template that can be instantiated.

Examples of cases where generic arguments can be suitable include statistical tests and calibration arguments.

One possible structured argument template is based on the appropriateness of each stage of the overall CoSMoS Simulation Project (92), essentially capturing the argument that:

- the appropriate instrument has been built: arguing over the Domain Model (128) science
- the instrument has been built appropriately: arguing over the Platform Model (149) design and Simulation Platform (161) implementation
- the instrument has been used appropriately: arguing over the Simulation Experiment (177)
- the results have been interpreted appropriately: arguing over the Simulation Behaviours (179) and Domain Behaviours (135)

This structure can be captured as a generic argument template:

[**claim** 1] the CoSMoS Simulation Project simulation is fit for purpose

   [**context** 1.1] Simulation Purpose documents role and criticality

   [**strategy** 1.2] argue over (i) the scientific domain, (ii) the implementation, (iii) the experiments, (iv) the interpretation of the results

   [**claim** 1.2.1] the Domain Model adequately captures the Domain for the Simulation Purpose

      [**strategy** 1.2.1.1] argue over the model content and assumptions

      [**context** 1.2.1.1.1] the documented assumptions

      [**justification** 1.2.1.1.2] sign-off from the relevant stakeholders

      [**claim** 1.2.1.1.3...] <project specific subclaims> ◇

   [**claim** 1.2.2] implementation: the implementation adequately captures the Domain Model for the simulation purpose

      [**strategy** 1.2.2.1] argue over (i) the derivation of the Platform Model from the Domain Model, and (ii) software engineering, testing, and calibration of the Simulation Platform

      [**claim** 1.2.2.1.1] derivation: the Platform Model is adequately derived from the Domain Model ◇

      [**claim** 1.2.2.1.2] software engineering: the Simulation Platform is adequately engineered ◇

   [**claim** 1.2.3] experiments: the Simulation Experiment is adequately performed

      [**strategy** 1.2.3.1] argue over use within Calibration, random seeds, analysis within the Results Model, and comparison with Domain results

      [**claim** 1.2.3.1.1...] <project specific subclaims> ◇

   [**claim** 1.2.4] interpretation of results: the Simulation Behaviours are adequately related to the Domain Behaviours ◇

Each undeveloped claim in this template needs to be developed with an argument strategy relevant to the specific simulation development.

Use Generic Argument

Intent

Instantiate a generic argument for use in a specific context.

Summary

- check that all generic elements are correctly instantiated
- check that Argument Context (196) is an accurate expression of the context of the instantiated Claim (192) and Strategy (194) components, and modify as appropriate
- check that Assumption (197) and Justification (198) annotations are appropriate and sufficient in the instantiated form, and modify as appropriate

Context

Argument Structuring Notation (190)

Discussion

When a generic argument structure meets the needs of a particular argument or part of an argument, it can be inserted, instantiated, and perhaps modified, to express the particular argument.

There are particular obligations when we use a generic argument, to ensure that the annotations and details of the generic argument hold for each specific instantiation.

Figure 8.9 shows a generic argument over the domain modelling, the software engineering, and the simulation results, instantiated for the prostate cell modelling example. An argument using this approach is most easily constructed retrospectively, or at least after the domain modelling phase, because it assumes a broad overview of the project development.

A generic argument may not be very useful early in development, because not enough is known to instantiate and extend it.

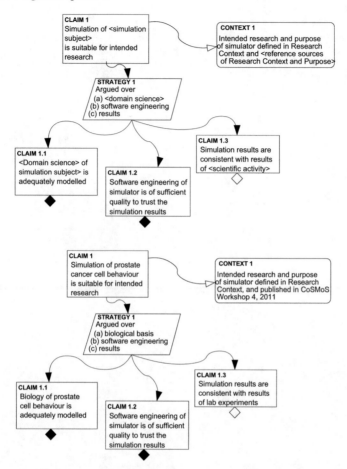

Fig. 8.9 The instantiation of a generic argument over domain, software engineering and results. It shows a possible two top levels of claim in such an argument (top) in generic form; (bottom) in instantiated form for the prostate model. See Argument Structuring Notation (190) and its composite patterns for explanation of notations

Part III
The CoSMoS Helper Patterns

This part presents "helper" patterns. These are more specific patterns of use only in certain domains, for certain purposes, or with certain modelling and implementation approaches.

This is not an exhaustive collection of patterns relevant to building scientific simulations. There will be other patterns specific to your domain. There are many existing software engineering patterns, some of which are referenced in the text and the endnotes.

Chapter 9
Modelling and Documentation Patterns

Abstract — In which we describe patterns covering particular modelling and documentation techniques.

9.1 Catalogue of patterns

Modelling approaches	
Agent Based Modelling (208)	Model the system in terms of agents, their behaviours, and their interactions.
Environment Orientation (210)	Mediate all inter-agent communication through the environments within which they co-exist.
Hybrid Model (212)	Use the most appropriate Modelling Approach for each part of the model.
Prototype (213)	Build an experimental model to explore or test ideas.
Seamless Development (214)	Use a consistent Modelling Approach across Domain Model, Platform Model, and Simulation Platform, to make validation arguments simpler and clearer.
Shortcuts (215)	Hard-code non-central mechanisms.

© Springer Nature Switzerland AG 2018
S. Stepney, F.A.C. Polack, *Engineering Simulations as Scientific Instruments: A Pattern Language*, https://doi.org/10.1007/978-3-030-01938-9_9

Communication and documentation

Ask [Silly] Questions (217)	Ask questions to improve understanding, even if you think the questions will seem trivial to the other person.
Take Notes (218)	Provide a record of key events.
Project Repository (219)	Use a project-wide repository to coordinate the project information.
Version Control (221)	Use a version control system for code, models, data, and documentation.
ODD Protocol (222)	Publish the simulation details conforming to the ODD protocol.

9.2 Modelling approaches

This section presents some specific examples of a possible Modelling Approach (111). Other approaches are possible; what is important is that they are chosen for a good fit with the Research Context (119), the Simulation Purpose (121), and the structure of the Domain (123).

Agent Based Modelling

Intent

Model the system in terms of agents, their behaviours, and their interactions.

Summary

- identify components and model as individual agents, with state and behaviour
- identify and model interactions between agents
- identify and model the environment, and how agents interact with it

Context

One possible choice of Modelling Approach (111).

Discussion

Agent Based Modelling (ABM) is a standard and well-established computational modelling technique. It provides a model expressed in terms of a population of *agents*, which can have *state* and *behaviour*, and which interact with other agents and their *environment*. This approach is also known as multi-agent simulation, and as Individual-based modelling[1] (IBM) [105].

The agent may be an atom or molecule, it may be a biological cell, it may be an organism, it may be a human agent, it may be an artefact such as a road vehicle or a building, it may be a star or a galaxy. Agents may act, react, and interact; their interactions may be determined by laws of physics or biology, or they may be governed by the rules used by intentional agents.

Agents more complex than atoms or small molecules have internal *state*. This can be modelled using state diagrams. These are the individual behaviours that will be implemented in the simulation.

Interactions between agents can be modelled using, for example, activity diagrams or Petri nets. These are the interaction behaviours that will be implemented in the simulation.

Growth – a change in the number of agents – can be modelled using growth grammars such as L-systems, or using Petri nets.

Some ABMs may comprise only agents. However, many will have a (non-agent-based) environment. It too must be simulated, and hence modelled. The choice of whether something is an agent or part of the environment is a modelling decision.

ABMs can exhibit high level system, or emergent, properties. These are typically the kind of behaviours seen in the Domain (123), and hypothesised to be emergent consequences of the agent behaviours. These properties should *not* be implemented in the simulation, but captured in the Domain Behaviours (135) model. If the hypotheses are correct, simulated analogues of these behaviours should emerge.

More on the background and philosophy of ABM can be found in [33]. Examples of ABMs abound. In biology, they are used to model everything from cells and bacteria, stigmergic systems, to ecology and population dynamics. In the social sciences, they are used to model cities [23], for simple world models such as Sugarscape [66], and as a general description of technique [90].

The CoSMoS approach is modelling language and implementation language independent. It can be used with ABM libraries and frameworks such as NetLogo, MASON or FLAME[2].

Related patterns

Environment Orientation (210) is a way of modelling agent interactions mediated by their environment.

Environment Orientation

Intent

Mediate all inter-agent communication through the environments within which they co-exist.

Summary

- represent all agent communication as being mediated by the environment
- move environment-specific information (for example, location, orientation, local chemical concentration levels) from agents into the environment
- as the environment mediates communication, use it to for a simulation of real world physics
- allow agents to exist simultaneously in multiple environments, thereby allowing a rich world to be simulated
- allow agents to merge information from their multiple environments into a complete picture (from their point of view) of the world and other agents

Context

A possible choice of Modelling Approach (111), as a variant of Agent Based Modelling (208).

Discussion

In standard Agent Based Modelling, there is frequent reference to communication between agents. However, communication between Domain agents is mediated their environment. For example, one flocking bird sees another because the former receives photons from the environment, which has transmitted them from the seen bird. The environment controls the direction of the photons (they might be refracted because of the medium they are in), how many there are (taking account of the inverse square relationship with distance, and atmospheric absorption), etc.

When using Environment Orientation, all agent communication is mediated by the environment. Agents deposit information into the environment, and retrieve it from the environment.

Agents might be able to communicate simultaneously in multiple environments. For example, flocking birds might communicate both by light and by sound. These separate environments might have very different effects on the communications they mediate; that for light could relatively easy transfer information about direction, while that for sound, especially low frequency sound, only transfers partial direction information.

When multiple environments are used to mediate communications, explicit consideration is given to merging the information, in the context of each agent, from those environments. The information in each environment is expressed in the context of a particular accuracy and resolution in the dimensions of the environment; for example, some environments might be 2 dimensional, some 3 dimensional. When combining this information the agent must use the multiple observations to build up a more complete "picture" of the complete world.

For more information on Environment Orientation, see [120].

Related patterns

Implementation Approach (160): Full environment orientation may not be compatible with the way in which some Agent Based Modelling libraries support agents; for example, they may require location to be an inherent characteristic of agents, or may be unable to support more than one environmental model, or may not support processing at the environment level.

Hybrid Model

Intent

Use the most appropriate Modelling Approach (111) for each part of the model.

Summary

- choose an appropriate Modelling Approach (111) for each of the various components and aspects of the system
- define in a rigorous manner how the approaches combine

Context

When choosing a Modelling Approach, if a single notation cannot express particular concepts without significant change or extension, or if it leads to unwieldy or difficult-to-read models.

Discussion

A hybrid model uses an appropriate kind of Modelling Approach for each aspect of the system.

When using more than one notation or modelling approach, it is important to record how the approaches are integrated: are there common modelling concepts between the approaches? does a stimulus from one kind of model form an input in another kind of model?

Hybrid models may be intra-model (more than one kind of model at the same stage of the CoSMoS process) or inter-model (different kind of models at different stages of the process). In our prostate running example (Chapter 3), we use Petri nets to model a reactive system with state machines to model transitional aspects (intra-hybrid); how these models formally combine and relate is defined in [63]. The Platform Model (149) for a Java based implementation would then require us to derive a class diagram (inter-hybrid).

Other examples of hybrid models include a model of plant growth using L-systems combined with an ODE formulation of hormone transport [183], and a combination of Geographic Information Systems (GIS) and ABM for simulating social and ecological processes [91].

Related patterns

A hybrid model can be seen as a form of Domain Specific Language (228), in which the semantics of one or more notations are modified and linked to form a coherent variant of the original notations that can cleanly express the assorted characteristics (e.g. reactive, transitional, structural, behavioural) of the domain or platform.

The parts of a hybrid model might be linked through a Metamodel (234).

Different models might be appropriate for different domains in a Multi-domain Simulation (242), or at different scales in a Multi-scale Simulation (244).

Prototype

Intent

Build an experimental model to explore or test ideas.

Summary

- develop a prototype, or sketch, of the problem and proposed solution
- factor the information discovered into the relevant models
- throw the prototype away and start again from what has been learnt; do not use it as the basis of a Multi-increment Simulation (240)

Context

During the Discovery Phase (95) and Development Phase (96).

Discussion

Modelling is a non-trivial activity, and there may well be false starts and blind alleys. Sometimes there are several options, and the best one can be found only by trying them out and comparing them. Some experimentation can be done without using the same level of rigour needed for the full development. But it must be recognised that these experimental models are fit only for the purpose of exploration, not as part of the final system.

So such experiments should be done as explicit prototypes, with the intention of being thrown away once they have served their purpose.

A Prototype can be used for many purposes, such as deciding on a modelling approach, scoping the domain model extent, understanding resource requirements, testing platform architectures, and structuring arguments.

Related patterns

Controlled prototyping contrasts with *Tweaking Code* (183).

Seamless Development

Intent

Use a consistent Modelling Approach (111) across Domain Model (128), Platform Model (149), and Simulation Platform (161), to make validation arguments simpler and clearer.

Summary

Choose notations that are as conceptually similar as possible to express the various models.

Context

A component of the Domain Model (128), Platform Model (149) and Simulation Platform (161).

Discussion

A system that had a Domain Model expressed as ODEs, a Platform Model expressed in UML, and a Simulation Platform implemented in a functional programming language would require considerable argumentation to ensure nothing had been lost or gained in transitioning between these.

Having a single conceptual notation running through (for example, UML for both Domain Model and Platform Model, and an object oriented implementation language for the Simulation Platform), reduces the risk of introducing errors in translation, and makes the argumentation easier.

However, model clarity should not be sacrificed for this purpose. In our prostate running example (Chapter 3), the Domain Model uses Petri nets to model a reactive system with state machines to model transitional aspects; how these models formally combine and relate is defined in [63].

There is more to Seamless Development than using the same notation. The structure of the models should also be as close as possible. For example, even if expressed in the same notation, a Domain Model and the Simulation Platform might potentially have very different concurrency structures, requiring considerable argumentation to demonstrate correspondence.

If the seamlessness has to be broken somewhere between the Domain Model and the Simulation Platform implementation (because their notations are quite different), then it is preferable to have as much seamlessness as possible between the Platform Model and the Simulation Platform (their respective notations as close as possible; for example, UML and an OO programming language). The closer to the implementation, the more low-level detail is in the models, and the harder it is to argue fitness for purpose. Therefore the link should be kept as straightforward as possible. Breaking seamlessness at the more abstract level, between Domain Model and Platform Model, should be relatively easier to argue valid.

Related patterns

Use a Domain Specific Language (228), to 'compile' from one notation to another.

Contrast Hybrid Model (212).

Shortcuts

Intent

Hard-code non-central mechanisms.

Summary

- decide which properties should be emergent, and which mechanisms can be hard-coded.
- Document Assumptions (108) to justify these decisions

Context

Scoping decisions related to the Research Context (119), and design decisions when moving from Domain Model (128) to Platform Model (149).

Discussion

Certain components need to be added to the Platform Model (149), particularly to simulate features of the domain taken as "given", such as the underlying "physics" of the Domain (123). Some of these might be included in the Domain Model (128) as properties of the system, but in the Platform Model, a decision needs to be made on how to simulate them.

Computational resources are limited, and not everything can be simulated in detail. It is necessary to hard-code some properties and laws at certain levels, rather than letting them all emerge from lower levels.

Some hard-coding is needed to implement features of the lower levels not being investigated, but needed in the simulation. These layers are sometimes called the "physics" of the simulation. These may include, for example, transport and movement mechanisms, where agents are moved by the simulation, rather than that movement emerging from lower level mechanisms.

Some hard-coding is needed even at the examined (emergent) levels, in order to provide some of the behaviours at those levels. This may include, for example, reproduction: the simulation makes new agents, rather than their creation being the result of some emergent mechanisms.

These hard-coded mechanisms are dubbed "shortcuts" [22]. Proxies and surrogates are also a form of shortcut.

It is important to ensure that any hard-coded mechanisms do not then "code in" any desired emergent properties. It is therefore important to Document Assumptions (108), paying particular care to examine the consequences of what conclusions can be drawn from the simulations results, and to justify why the chosen hard-coding will not invalidate the Simulation Purpose (121).

Related patterns

Environment Orientation (210) is a pattern for shortcutting a lower level "physics" transport layer.

Ensure that any shortcut does not *Program In the Answer* (170).

9.3 Communication and documentation

Good communication and documentation practices are crucial to a successful simulation project. Cross-disciplinary work involves much communication between the disciplines, and much unfamiliar material needs to be absorbed and understood. Domain (123) issues well-understood to the Domain Scientist (101) need to be communicated to the Domain Modeller (103), sufficient for building appropriate models. Software issues well-understood by the Simulation Engineer (105) need to be communicated to the Domain Scientist (101), sufficient for understanding and agreeing simulation compromises. The Argument Modeller (106) needs to communicate what evidence is needed to build a convincing argument to the entire team.

Good documentation strategies for software engineering [196] can be adapted for the full simulation project. Multiple aspects require some form of documentation:

- meetings between project members
- models
- links between models
- software
- results and analysis
- design decisions: their justifications and consequences
- problems and issues to be resolved
- argumentation

In addition to these standard issues, some CoSMoS-specific patterns are noted here.

Ask [Silly] Questions

Intent

Ask questions to improve understanding, even if you think the questions will seem trivial to the other person.

Summary

- take asking questions, and answering them, seriously
- use the opportunity when defining core terms, especially for the Glossary (127)
- use the opportunity to challenge assumptions

Context

When people with different Roles (99) are interacting.

Discussion

The Simulation Engineer (105) and the Domain Scientist (101) have different background and expertises. What may seem obvious to one is often not at all obvious to the other. Non-experts may have naive misunderstandings of the subject. However, people are often reluctant to ask what they think might be perceived as a naive question.

The team needs to work together to understand assumptions and limitations. It is important to culture an environment in the project where people feel able to ask such questions. Sometimes an apparently naive question can lead the expert to re-evaluate their understanding: the question is being asked from a different perspective, and might highlight a tacit assumption.

As questioner, you should ask follow-up questions to test your understanding of the answers. As answerer, you should monitor the follow-up questions; if questions continue when you think they have been answered, or go off in an unexpected direction that might indicate a misunderstanding, probe what the misunderstanding is, and become the questioner yourself: "why did you ask that question?"

Related patterns

Trust between the Roles (99) should foster an atmosphere where questions can be asked and answered seriously.

Beware the blind trust of the *Uncritical Domain Scientist* (143) or *Uncritical Domain Modeller* (143): you are not asking enough questions.

Take Notes

Intent

Provide a record of key events.

Summary

- appoint a Scribe for each meeting
- make a note of all the relevant issues

- store notes for future reference

Context

When people with different Roles (99) are interacting.

Discussion

A CoSMoS Simulation Project (92) involves a lot of facts, definitions, data and information sources, decisions, arguments, and more. It is crucial to keep a record of these, both for future reference, and to help new members joining the team, particularly when revisiting assumptions, and for a Multi-increment Simulation (240).

Take notes in meetings between members with different Roles. These can include paper notes, photographs of whiteboard diagrams, links to information, and more. In some cases, it may be appropriate to record meetings, although the overhead of transcribing, or searching, a recording can be high.

One of the useful sub-Roles is that of Scribe: this should be assigned, or volunteered, for each meeting. This role's responsibility is to ensure all the key points are captured, stored in the Project Repository (219) for future reference, and summarised to allow absent members to catch up.

Project Repository

Intent

Use a project-wide repository to coordinate the project information.

Summary

- set up a project-wide repository
- include a Version Control (221) system
- decide what aspects of the project to document, and what to put under version control
- store all relevant information in the repository
- make the repository accessible to all project members
- decide what parts of the repository should have public access

Context

Setting up a CoSMoS Simulation Project (92); a component of the Research Context (119) pattern.

Discussion

It is essential to organise and store the artefacts (meeting notes, models, diagrams, arguments, code) of a CoSMoS Simulation Project (92).

The Project Repository should be planned and used to store all project documentation in a form that is accessible to all project members. Depending on the size of the project, and background of the project members, this repository may be, for example, a shared directory, a shared wiki, or a generic repository. Include the use of a Version Control (221) system to ensure that different versions of models, code, data, arguments, and other documentation are synchronised as appropriate.

A shared storage structure (such as a shared drive or a wiki) provides a fluid and agile means of holding documents and records of meetings. Some shared spaces can be customised to a particular structure defined in your House Style (229). A good shared space supports collaborative effort, and the detailed structure develops and emerges naturally [152]. There are also shared storage structures with associated project management functions (for example, Trello). A shared space with logging to allow changes to be tracked, and rolled back if necessary, is useful.

Version controlled repositories such as GitHub, or bitBucket, originally designed for code, can be used to store and version-control all project documents along with code. Such repositories are usually suitable for a large project. However, repositories may be seen as hard to access by less technical members of the team; it is important that all project team members can easily enter and access all project documentation.

Large simulation projects need more than just an appropriate shared storage: formal documentation processes are needed to ensure that all material is appropriately stored and that the relevant project members are notified of changes.

Design your repository with suitable access controls. If you are engaging in an Open Science project (see p.40), the entire Project Repository could be made publicly visible. A simulation associated with commercial development may be allowed to make only certain documents publicly available.

Related patterns

Document the approach taken in your House Style (229).

Take Notes (218), and store them in the Project Repository.

Version Control

Intent

Use a version control system for code, models, data, and documentation.

Context

A component of the Project Repository (219) pattern.

Discussion

A CoSMoS Simulation Project generates many artefacts: arguments; models; code; experiments including parameter values, input data, random seeds, and results. These artefacts are generated iteratively, and in parallel. It is important to keep these under appropriate version control in a suitable repository, to allow the whole team access to the latest version, and to ensure that previous versions can be accessed as required.

In particular, it is important to have access to the correct version of documentation, models, data, software, results, and analysis used to produce particular public artefacts, such as published papers, for transparency and reproducibility.

Related patterns

Beware *Tweaking Code* (183) and failing to make an updated version.

Version control is essential in order to Propagate Changes (168) methodically. Document the approach taken in your House Style (229).

A Multi-increment Simulation (240) will need especially careful version control to separate out the results and understanding gained in each increment.

ODD Protocol

Intent

Publish the simulation details conforming to the ODD protocol.

Summary

Extract ODD protocol information from the CoSMoS documentation.

Context

Publishing your simulation description in a standard format.

Discussion

The ODD (Overview, Design concepts, Details) protocol [103, 104] has been devised as a standard way of describing simulation models (particularly in the ecological domain), specifically to aid reproducibility of implementation.

A CoSMoS Simulation Project (92) contains all the information needed to give an ODD protocol description. The ODD protocol information is only part of the entire CoSMoS model set: it is mostly contained in the Domain Behaviours (135), Platform Model (149), Data Dictionary (132), and the Research Context (119). It explicitly does not cover Results Model (174) features of Simulation Experiment (177) and Sensitivity Analysis (175): it considers these to be the "methods" part of a description; the ODD protocol covers only the "materials" part of a standard scientific article [103]. Also, it is more concerned with the implemented model, so does not distinguish the concepts of Domain Model (128) and Platform Model (149). "Although the protocol was designed for ABMs, it can help with documenting any large, complex model" [104].

In addition to the explicit material noted in the table below, the ODD protocol also recommends making the source code (that is, the Simulation Platform (161)), available, and using code comments to highlight the various parts of the protocol information [103].

In the table below, the thumbnail description of each ODD protocol component is abstracted from [103–105, 184], which contain fuller descriptions and explanations of the purpose and role of each component. The headings follow the revised protocol in [104]. This table outlines where the ODD protocol information appears in a CoSMoS model; for more detail see [204].

ODD heading	ODD component	CoSMoS location
purpose	why the model was developed; what is to be used for	Simulation Purpose (121)
entities, state variables and scales	"what the model world is"	
entities	the different types of agent, including the environment; any hierarchical collective structure of agents (cells, organs, organisms; individuals, communities; etc.) that have their own identity and behaviour	Platform Model (149) class diagram: relationships between agent classes
state variables	the agents' state variables or attributes (characterising individual properties, parameter sets, strategy or behavioural choices, etc.), including units	Platform Model (149) class diagram: agent instance variables
scales	lengthscales (for a spatial model; for example, grid size and number of grid cells) and timescales (what real time a timestep represents; the number of timesteps)	Data Dictionary (132), Space Model (155), Time Model (158)

process overview and scheduling	"how the model world changes": what the agents do (growth, movement, reproduction, etc.) and their in what order (the order in which state variables are updated, synchronous or asynchronous, concurrency); how time is modelled (discrete v continuous)	Platform Model (149) components expressed as activity diagrams and state diagrams, or as pseudocode (as recommended in [104])
design concepts	(see [184],[105, ch.5])	
basic principles	concepts, theories, hypotheses and modelling approaches underlying the design	Research Context (119) and Domain (123) (general concepts); Domain Model (128) (theories and hypotheses); Modelling Approach (111)
emergence	system level phenomena that emerge from the individual behaviours, rather than being imposed	Domain Behaviours (135), not carried over to Platform Model (149)
adaptation	rules for making decisions and changing behaviour	relevant Domain Model (128) entity behaviours
objectives	measures of an agent's success (fitness, utility); criteria agents use to achieve this	relevant Domain Model (128) entity behaviours
learning	how agents change their adaptive traits	relevant Domain Model (128) entity behaviours
prediction	how an agent predicts future consequences in order to make decisions	relevant Domain Model (128) entity behaviours

sensing	what state variable values (of itself, and of the environment) does an agent have access to, to guide or influence its behaviour	Domain Model (128) entity sensors; Platform Model (149) class diagram, relationships; Environment Orientation (210) model (if used)
interaction	direct and indirect interactions between agents; representation of communications	Domain Model (128) entity interactions; Platform Model (149) class diagram, relationships; Environment Orientation (210) model (if used)
stochasticity	what real world variability is modelled by randomness	Stochasticity Model (154)
collectives	collections of agents that behave as entities in their own right; which are emergent, and which are explicitly modelled	emergent entities in Domain Behaviours (135); explicitly modelled collections in Platform Model (149), often as Shortcuts (215)
observation	what data is collected from a simulation run for analysis	Simulation Experiment (177) and Results Model (174) entries in Data Dictionary (132)

initialisation	initial state of simulation at $t = 0$ (number and state of agents and environment)	Data Dictionary (132) for Simulation Experiment (177)
input data	data that drives environmental variables (such as rainfall or harvesting regimes) [103], imported from external files or models	Data Dictionary (132) for Simulation Experiment (177)
submodels	"how the model world changes, in detail": mathematical equations, rules, parameters that define the processes	Platform Model (149) implementations of Domain Model (128) definitions

The documentation resulting from a CoSMoS Simulation Project (92) contains places for all the information needed to document an ABM using the ODD protocol. Hence CoSMoS can be used to develop ODD-compliant simulations. However, the material is scattered through several CoSMoS artefacts. If the ODD protocol is to be used to present the results of a CoSMoS-developed Simulation Experiment (177), then it would be sensible to devise a House Style (229) and Project Repository (219) structure that specifically tag and make accessible the ODD-relevant information in each of the CoSMoS artefacts.

Chapter 10
Real world simulation patterns

Abstract — In which we describe patterns for real world issues: large scale, messy, incremental developments; composing simulations; using metamodels.

10.1 Catalogue of patterns

Development process	
Domain Specific Language (228)	Use a special purpose small language to capture domain structures, to aid communication and ease development.
House Style (229)	Define a house style for specific use of CoSMoS.
Partial Process (230)	Use only part of the overall CoSMoS approach.
Post Hoc (232)	Reverse engineer the CoSMoS artefacts.
Refactor (233)	Modify a model, code, or argument to improve its structure, without changing its meaning.
Metamodel (234)	Build a model of a model.

© Springer Nature Switzerland AG 2018
S. Stepney, F.A.C. Polack, *Engineering Simulations as Scientific Instruments: A Pattern Language*, https://doi.org/10.1007/978-3-030-01938-9_10

Engineered systems

Engineered Domain (238)	Build a simulation of an engineered rather than natural domain.
Embodied Simulation (239)	Embody a Simulation Platform in a robot, or in a robot swarm.

Large scale development

Multi-increment Simulation (240)	Develop the project in small manageable increments.
Multi-domain Simulation (242)	Develop a project that encompasses multiple distinct domains.
Multi-scale Simulation (244)	Develop a project that encompasses multiple distinct levels or scales.

10.2 Development process

Domain Specific Language

Intent

Use a special purpose small language to capture domain structures, to aid communication and ease development.

Summary

Choose an appropriate existing DSL, or (for experts only) design and use a new DSL specific to the project.

Context

When constructing the Domain Model (128).

Discussion

By Domain Specific Language (DSL) we here mean a modelling language defined for use in a specific Domain. A DSL may be more appropriate for building all or part of the Domain Model than a general purpose language such as UML, if it is tailored to capturing the particular domain properties in a more natural manner. Thus use can ease communication of the model to the Domain Scientist (101).

Existing DSLs can be used or modified, such as the combination of Petri nets and state diagrams used in the case study presented in Chapter 3. New DSLs can be developed [78], but this is a non-trivial task, and should be undertaken with care.

DSLs are common in computer programming[1], developed to aid specific computational tasks; they include such languages as LATEX for document preparation, TikZ for creating graphics, and awk for data manipulation. Specific domains may have their own DSLs; for example, SBML (Systems Biology Markup Language)[2], can be used to represent models in a machine-readable format.

Human-readable modelling languages are needed for communication at the domain model level, and so tend to be diagrammatic. There needs to be a careful trade-off between being so informal that the DSL is actually just a Cartoon (124) with no well-defined meaning, and being so formal that the language becomes so swamped with detail that it becomes hard to use, and makes the models hard to understand.

Related patterns

Use a Metamodel (234) to define the concepts in the DSL.

Argument Structuring Notation (190) is a DSL for presenting fitness-for-purpose arguments.

Use of a DSL works against Seamless Development (214), unless the DSL can be used throughout the development.

House Style

Intent

Define a house style for specific use of CoSMoS.

Summary

- instantiate certain patterns with specific cases
- document variant use of patterns
- add new patterns
- revisit the House Style after each CoSMoS Simulation Project (92)

Context

During and after completion of a CoSMoS Simulation Project (92)

Discussion

After having developed several simulations using CoSMoS, you discover you have made some repeating decisions, and variations to the process. For example, you might have a common structure for the Project Repository (219), use a particular Version Control (221) system, use a particular Modelling Approach (111) and Implementation Approach (160), omit certain steps, or include extra steps that are necessary in your research domain.

Document these decisions and variations of the standard CoSMoS approach as modified or new patterns, to form your House Style. This will simplify future projects, and provide guidance for new project members. Revisit and update your House Style after the completion of each project, to keep it up-to-date, and to gain the most from your experience.

These modified and new patterns should include a step to check that the documented decision is still valid for the new project, to avoid *One Size Fits All* (142).

Partial Process

Intent

Use only part of the overall CoSMoS approach.

Summary

Use the Research Context (119) to tailor which parts of the overall CoSMoS approach to exploit.

Context

An approach to the overall CoSMoS Simulation Project (92)

Discussion

Sometimes there is no need, or no time, to carry out an entire CoSMoS Simulation Project (92). In such cases, use the Research Context (119), and if appropriate the Simulation Purpose (121), to determine what parts of the approach to exploit, and what parts can be safely dropped. Different circumstances will result in different tailorings. (See [150] for a more detailed discussion of tailoring a development process.) Examples include:

- **Part of the domain**. Particularly when the domain is large and complex, you might choose to focus effort on a subdomain, in order to understand that in more depth, and mimic the remainder using 'proxy' inputs, such as from experimental data, or user input. In socio-technical systems, simulating just the technical part, and recruiting typical users to provide a 'live' model of the socio part, can alleviate problems of simulating intelligent reflective actors.
- **Model only**. Sometimes all that is needed is a domain model. This might provide sufficient insight so that a simulation is not necessary, or it might demonstrate that the domain is not yet well-enough characterised for building a simulation.
- The above tailorings may be combined, where parts of the domain are modelled only, and other parts are also simulated. In this case, the model can help guide the simulation experiments, or coordinate multiple subdomain simulations. See the example discussed in Multi-scale Simulation (244).
- **No Argumentation**. Due to resource limitations, you may decide not to Argue Instrument Fit For Purpose (186). If so, at least this decision should be argued. And if so, do not neglect to Document Assumptions (108): these are a crucial part of the modelling process, and invaluable when you come to build on the project later.

Other tailorings are possible, depending on the Research Context (119).

Related patterns

A tailoring may be Post Hoc (232), starting from artefacts not developed using the CoSMoS approach.

If you use the same tailoring on multiple CoSMoS Simulation Projects, add it to your House Style (229).

Post Hoc

Intent

Reverse engineer the CoSMoS artefacts.

Summary

Tailor the CoSMoS approach to develop the required artefacts in an appropriate order.

Context

An approach to the overall CoSMoS Simulation Project (92)

Discussion

A CoSMoS Simulation Project states which core artefacts (models, code, arguments, etc) are needed, but these do not need to be produced in a specific order. The standard approach is to proceed through Discovery Phase (95) to Development Phase (96), to Exploration Phase (97), but other orders are possible. The most obvious of these choices is whether to Argue Instrument Fit For Purpose (186) during or after the rest of the development.

Another case where artefacts may be developed in a different order is when a Simulation Platform (161) exists that was not developed using the CoSMoS approach. If this is to be used as the basis of a further development using the CoSMoS approach, reverse engineer a Platform Model (149) and Domain Model (128) from the Simulation Platform code and any design documentation available. This will almost certainly uncover undocumented assumptions, and may uncover hard-coding of Domain Behaviours (135). When reverse engineering in this way, rather than faithfully replicate exactly what is in the code, develop an appropriate Domain Model and Platform Model, with the agreement of the Domain Scientist (101). Then amend the Simulation Platform to be consistent with these agreed models, and, if appropriate, rerun each Simulation Experiment (177) to determine the effect of the code changes on previous results. These models and the amended Simulation Platform are then suitable artefacts to form the baseline for the next increment of a Multi-increment Simulation (240).

Not all reverse engineering is done to develop a new increment. It may be used to uncover and analyse the assumptions and models in a given simulation, and to reimplement an existing system. This approach is used in [1] to analyse Schelling's model of segregation.

Andrews and Stepney [12] report the reverse-engineering of a Platform Model and Domain Model for Aevol, a sophisticated artificial evolution simulator. The aim in that case was to extract a Metamodel (234), in order to develop a different, but related, simulator.

Other reverse-engineering approaches can be tailored, depending on the Research Context (119).

Related patterns

A tailoring may be a Partial Process (230), not developing all the core artefacts.

Refactor

Intent

Modify a model, code, or argument to improve its structure, without changing its meaning.

Summary

Change the structure of a CoSMoS artefact without changing its meaning, to improve its structure, or to support a proposed subsequent modification.

Context

Any time during the CoSMoS Simulation Project (92) when an artefact (model, code, argument) needs to be modified.

Discussion

When engaged in CoSMoS Simulation Project (92), there will be times when changes are needed: errors may be discovered; new concepts or functionality may need to be added; features may need to be removed. Make these changes in a controlled manner, across the various interconnected models and arguments.

Make changes incrementally, to help control complexity. Changes can be classified as changes of *meaning* (fixing a bug, adding functionality), and changes of *structure* (modifying the component without changing its meaning, so that a subsequent change of meaning can be made more easily). To change structure without changing meaning, Refactor (233) the artefact, be it glossary, models, experiments, code, or arguments.

Propagate the refactoring change through the project, to ensure all the artefacts are consistent. For example, a concept might be renamed: propagate the new name through the Glossary (127), the models where it is used, the code where it is implemented, and the arguments where it is referenced. Make several small changes, propagating each one fully before doing the next, rather than one large change, to reduce the chance of introducing errors.

Code refactoring is discussed in [75], and refactoring code to patterns in [137]. These code ideas can be adapted to refactoring models, and especially refactoring them to patterns such as the ones that form the CoSMoS Simulation Project.

Related patterns

This pattern is a key part of Multi-increment Simulation (240), but can be used at any time a change is needed.

Metamodel

Intent

Build a model of a model.

Summary

- decide if an explicit metamodel is needed for the chosen Modelling Approach (111)
- design the concepts needed in the modelling approach
- design a semantics and syntax for the modelling approach, as appropriate

Context

- defining a Domain Specific Language (228)

- linking the Domain Model (128), Platform Model (149) and Results Model (174)
- linking the components in a Hybrid Model (212)
- linking the components in a Multi-domain Simulation (242)
- linking the levels in a Multi-scale Simulation (244)

Discussion

The CoSMoS Simulation Project (92) is based on the use of models. A model captures the relevant concepts in a particular language: the Domain Model captures the concepts in the domain of interest; the Platform Model captures the concepts to be implemented in the Simulation Platform (161) code. A Metamodel provides the analogous concepts for writing a model: it defines the kinds of things that can occur in the model (it is the model of the model) [142, Chap. 8].

A Metamodel may be implicit, semi-formal, or fully formal. Computational models are more likely to have formal metamodels.

Each Modelling Approach (111) has its own Metamodel. For example, in an Agent Based Modelling (208) of a system with emergent properties, the Metamodel could include concepts such as *Agent*, *Rule*, and *Emergent*. For a different style of model, such as an ODE model, the Metamodel could include different concepts, such as *Concentration* and *RateOfChange*.

When designing a new modelling approach, it can be helpful to make the metamodel explicit [22]. For example, when using or designing a Domain Specific Language (228), a Metamodel can be used to define the concepts of the DSL.

For example, consider the Metamodel for Petri nets used in prostate example (Chapter 3). A Petri net is a bipartite graph with two types of (named) nodes, places and transitions, joined by arcs. Places can hold tokens; tokens are produced and consumed in transitions. (Since the graph is actually a direct graph, we could refine this Metamodel further, for arcs entering and leaving the places and transitions; we choose not to do so here.) Note that there are many constraints not explicitly represented in this model: a token in a place was produced by a transition connected to that place by an arc; the graph is connected; names are unique; and so on. For a full DSL, these constraints would also need to be specified.

Fig. 10.1 Petri net
metamodel. Places and
Transitions have Names.
Each Arc joins one Place
and one Transition; Places
and Transitions may have
multiple Arcs. Each Token
is held in one Place; it was
the output produced by
one Transition, and can
be the input consumed by
one Transition. Transitions
can consume and produce
many Tokens

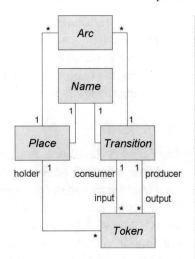

Consider the Metamodel for state diagrams used in prostate example. Again, there are many constraints not explicitly represented in
this model, for example, the need for unique names; that a transition
cannot join an entry node to an exit node.

Fig. 10.2 State diagram
metamodel. Entries, States,
and Exits all have Names.
States can have sub-states.
A Transition connects
Entry, State, and Exit
nodes. A Transition has
one source (Entry or State),
and one target (State or
Exit)

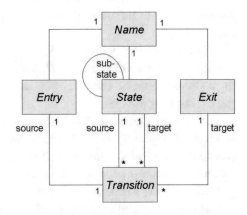

If we want to formally link the Petri net Metamodel and the State
diagram Metamodel for their use together as a Hybrid Model (212) in the
prostate example, we could note that the state diagram *Entry* and *Exit*
names are the same as Petri net Place *Name*s, but we would have to
rename one of the *Transition*s, as they are different concepts.

The CoSMoS Domain Model and Results Model share a common
Metamodel. This does not mean that the two models are identical; it

means that they are cast in the same language. The Domain Model has instances of Metamodel concepts that capture specific Domain concepts; the Results Model has instances of simulation analogues of those Domain concepts. So where an agent-based Domain Model may have a *Bird*, the Results Model will have a *Boid*, the simulation analogue of *Bird*; both are instances of the agent-based Metamodel concept *Agent*. The Results Model has *Data* instances, which stand in the same relation to its *Agent* instances as they do in the Domain Model (so if the Domain Model has bird positions and velocities, the Results Model has the corresponding boid positions and velocities). This allows a direct comparison of the models in Domain terms.

The Metamodel of the Platform Model is different from Metamodel of the Domain Model and Results Model. In particular, it has no *Emergent* concept; it is important that there is no way to program the desired answer into the simulation: it must emerge. (If the Research Context is not concerned with emergent properties, but some other kind of property, it is equally important to ensure that this other property does not get programmed in to the simulation.)

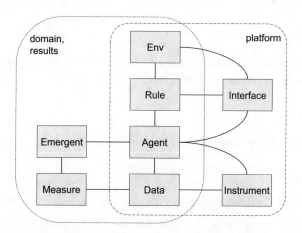

Fig. 10.3 A CoSMoS metamodel for agent-based simulation. The dashed boxes indicate the components of the metamodel that are used to describe the domain and results models, and the platform model. From [14, Fig.3]

Related patterns

The CoSMoS Metamodel demonstrates an approach to using a meta-model to formally link disparate models, as may be used in a Hybrid

Model (212), in a Multi-domain Simulation (242), or in a Multi-scale Simulation (244).

Seamless Development (214) is aided by using a Modelling Approach (111) and Implementation Approach (160) with related metamodels.

10.3 Engineered systems

Although most of the examples in this book are about modelling natural biological systems, CoSMoS can be applied to building a simulation of any system: physical, biological, social, engineered. Here we present two patterns appropriate for engineered systems.

Engineered Domain

Intent

Build a simulation of an engineered rather than natural domain.

Summary

Use the standard CoSMoS approach, but interpret a disagreement between results and domain differently.

Context

An approach to the overall CoSMoS Simulation Project (92).

Discussion

The standard CoSMoS Simulation Project (92) assumes that the Domain (123) is a natural system, and that the overarching Simulation Purpose (121) is to investigate hypotheses about that system, to gain deeper understanding. If a Simulation Experiment (177)'s results disagree with domain experiments or observations, this suggests there may be something missing or wrong with the Domain Model (128), and that it does not capture the Domain adequately.

But the Domain may be an engineered, or designed, system, and the Simulation Purpose be to provide an *in silico* analogue of the designed system. In such a case, the CoSMoS approach can still be used, but there is a different interpretation of the Results Model (174). The Domain Model is the engineering *specification* of the Domain. The Domain is *engineered* to respect the Domain Model. If a Simulation Experiment's results disagree with Domain behaviour, is suggests there may be something missing or wrong with the engineered Domain, and that it does not implement the domain model Domain Model.

Embodied Simulation

Intent

Embody a Simulation Platform (161) in a robot, or in a robot swarm.

Summary

Either (i) embody a simulation within a robot, providing that robot with an internal model; or (ii) construct a simulation with parallel interacting real world elements, i.e. robots.

Context

Embodied Simulation is a type of Simulation Platform (161). It is also a distinctive Modelling Approach (111).

Discussion

The field of robotics identifies two different kinds of embodied simulation.

Firstly, in the field of intelligent robots, specifically addressing the problem of machine consciousness [115], the notion of robots with internal models has emerged in recent years. Such a model is a simulation of either the robot itself, or its external environment (including other agents), or both – and the simulation is embedded within the robot; the simulation is literally embodied. Such a simulation might allow a robot to "try out" alternative sequences of motor actions, to find the sequence that best achieves the goal (for instance, picking up an object), before then executing that sequence for real. Feedback from the real world actions might then also be used to calibrate the robot's internal model. The robot's embodied simulation thus improves over time, i.e. the robot learns. If the purpose of the robot's embodied simulation is to model the actions of other agents (i.e. robots or humans), then the model could be said to represent an artificial Theory of Mind; an approach which not surprisingly implements the simulation theory of mind. For an example, see Holland's Cronos robot [154].

Secondly, in the field of swarm robotics [198] experimental work with the aim of developing or discovering algorithms and/or investigating emergent properties of swarm intelligent systems is typically carried out using a number of autonomous mobile robots in an instrumented arena. Each robot is programmed with a set of behaviours and the robots interact with each other and with their physical environment and, with appropriate design of the robots and their behaviours, we can observe desirable or interesting emergent, or self-organising, properties. The complete system is a simulation of swarm intelligence [61] with real robots; it is an embodied model of swarm intelligence, hence an embodied simulation. There are several advantages of such a system. It typically runs faster than the same system of robots modelled in computer simulation and obviates the need for a visualisation model. It is a truly parallel asynchronous system, so it preserves concurrency and avoids quantisation artefacts. Physics and noise come "for free", thus providing a natural implementation of stochasticity.

Related patterns

Visualisation Model (181)

10.4 Large scale development

Multi-increment Simulation

Intent

Develop the project in multiple small manageable increments.

Summary

Start with a small project covering part of the Domain (123). Incrementally increase the scope, until the full system has been developed.

Context

An approach to the overall CoSMoS Simulation Project (92)

Discussion

Depending on the scope of the Research Context (119), a CoSMoS Simulation Project can be a large undertaking. It is easy to get lost in the details. And for a research project, there may be many unknowns that need to be discovered along the way, and potential changes of direction as these are exposed, so the full development cannot be planned in detail from the start.

Use a Multi-increment Simulation approach to development. At each increment, develop a full system, but of reduced scope. For each new increment, decide what extra of the Domain (components, behaviours, experiments) is to be added or changed. This may require you to Refactor (233) the system, to a greater or lesser extent, to allow for these changes; take care to propagate changes throughout, to maintain a consistent project.

At the end of each increment, you will have a complete, albeit reduced functionality, system. This has advantages if funding is disrupted or patchy: there is always some system that can be used now, and upgraded later.

If a Simulation Experiment (177) indicates there are errors or unknowns in the Domain Model (128), or if the whole project is using an exploratory approach, this can lead to decisions to change the Research Context (119) or Simulation Purpose (121). An incremental approach reduces the effort expended to reach such a decision point, and allows a more responsive approach to new findings. It also helps keep the development focussed on what is needed now, for this increment, not on what is assumed to be needed in the future. These advantages typically outweigh the effort needed to Refactor (233) between increments.

In software engineering, such an incremental approach is known as "Agile development", and comes in various flavours [27, 161]. In such an agile approach to Multi-increment Simulation, it is essential to keep a close watch on the Research Context and Simulation Purpose. As experience with the system increases, new and unanticipated demands may be made of it [31]; it is important to ensure that new uses do not violate earlier assumptions and scoping.

During an increment you may generate interesting but out of current scope ideas. Document these ideas, so that they are not forgotten, tagged as **!! Future**. These form a "to don't" list (as opposed to a "to do" list): a list of ideas for future reference, in a manner that makes it clear they are not to be included in the current increment. Some of these ideas may also prompt the recognition of assumptions in the current increment: that it does not support these future properties [98].

Related patterns

Consider using a Prototype (213) to understand specific issues.

Revisit the Research Context (119) and Simulation Purpose (121), and associated models and arguments, between increments.

Use Version Control (221) to keep access to the models and results of each increment.

Multi-domain Simulation

Intent

Develop a project that encompasses multiple distinct domains.

Summary

- build a Domain Model (128) of each individual Domain (123)
- build a framework model of how the Domains interact
- use a Metamodel (234) to find common concepts

Context

An approach to the overall CoSMoS Simulation Project (92)

Discussion

Complex systems can span multiple domains: socio-technical systems, socio-ecological systems, eco-geo-climate systems, and more.

One approach is to model the individual Domains, and also explicitly model how they interact. With engineered systems, these interactions might be designed to flow through well-defined interfaces. With natural complex systems, interactions can be broad and messy. The modelling process should expose these issues.

Forrester and Greaves [73, 101] use a *framework model* to define the interactions between multiple Domain Models. They find that the framework model is necessary to understand the multiple Domain Models, but that in some cases joining the parts in a fully integrated Simulation Platform is not always needed; once the framework model is understood, how the individual component models fit into the big picture is clear. In more complex cases understanding the interacting dynamics will require an integrated simulation, however.

Use a Metamodel encompassing the various Domain Model (128)s, to capture the commonalities and differences; for example, see Figure 10.3.

Sometimes there may be two separate Domains, with one inspiring the other, but not connected to it explicitly. For example, in bio-inspired engineering, there is one biological Domain used to inspire the design of an engineered Domain. In such a case, there is no need for a framework model; the Metamodel can be used to help formalise the relationship between the two Domains [14], see Figure 10.4.

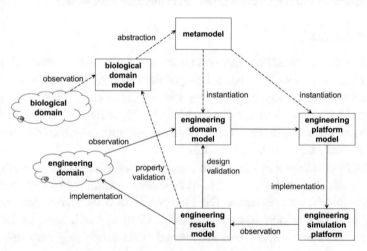

Fig. 10.4 A bio-inspired engineered domain. From [14, Fig.7]

Related patterns

Consider using a Prototype (213) to understand specific issues.

If the different Domains are modelled with different Modelling Approach (111)es, this is a Hybrid Model (212).

Particularly if there are more than two Domains, consider modelling the interactions of the domains through some medium, using Environment Orientation (210).

Multi-scale Simulation

Intent

Develop a project that encompasses multiple distinct levels or scales.

Summary

- build a Domain Model (128) of each individual scale/level
- build a framework model that links the separate Domain Models
- consider having both a detailed and a summary Domain Model (128) at lower levels
- use a Metamodel (234) to find common concepts

Context

An approach to the overall CoSMoS Simulation Project (92)

Discussion

Multi-scale simulation is non-trivial, and an active area of research [29, 53, 114, 132]. Here we give a few pointers to how CoSMoS's model-based approach may help. In this view, a multi-scale simulation is a special case of a Multi-domain Simulation (242), where the different Domains represent different scales, which constrains how the framework model links the individual models.

Multi-scale system have both multiple spatial scales, and multiple timescales. Lower levels tend to have shorter spatial scales and correspondingly faster timescales. Each level will need its own Space Model (155) and Time Model (158). The framework model will need to define how these spaces and times are related. Timebands [40] may help in building the framework's Time Model.

One major issue with multi-scale simulation is that the lower levels, with their finer granularities of space and time, can overwhelm the computational resources. Shortcuts (215) will be necessary. One approach is to use a detailed stand-alone lower level model, and use it to inform the design of a more abstract lower level counterpart that can be integrated into the overall system.

In our prostate running example (Chapter 3) is simple multi-scale model and Hybrid Model (212). It has a low level state diagram model to describe the internal behaviour of individual cells, and a Petri net model of the higher level cell population. How these models formally combine and relate is defined in [63].

Related patterns

Consider using a Prototype (213) to understand specific issues.

If the different levels are conceptualised as different domains, this is also a Multi-domain Simulation (242).

If the different levels are modelled with different approaches, this is a Hybrid Model (212).

Part IV
CellBranch Case Study

This part provides a fully worked CoSMoS Simulation Project. It is presented in a "warts and all" fashion, with only minor modifications to the original documentation, made to bring it in line with the current version of the patterns, and to correct minor errors. This provides a view of a real CoSMoS project, not of a 'perfect' one. You may wish to argue over the model, the decisions, the assumptions. That is in some sense the point: all is made visible to critique.

The presentation has two increments, but the project did not initially start as an incremental development: this was decided part way through the development. As a result, there is some contextual material presented in increment 1 that more logically fits in increment 2. We could Refactor (233) the documentation to separate out the material, but we leave it as originally developed, to illustrate a more realistic state of affairs. Some '!! **Future**' tags have been added to flag such material.

The full biological descriptions of the case study Domain have been maintained in this presentation, to illustrate a representative level of detail provided for publication of the documentation. However, many readers may wish to skip this detail, and focus on the use of the CoSMoS patterns.

The original versions of the case study documented in this Part have appeared in [98–100].

Acknowledgements

We gratefully acknowledge the funding of the CellBranch project by the UK's Biotechnology and Biological Sciences Research Council (BBSRC), project reference BB/L018705/1.

Chapter 11
Introduction to the CellBranch simulation

The principal barrier to gaining understanding of embryonic stem (ES) cell regulatory networks is their complexity. Reductionist approaches overlook much of the complexity inherent in these networks and treat the ES cell regulatory system as more or less equivalent to the sum of its component parts, studying them in relative isolation. However, as we learn more about regulatory components it becomes increasingly difficult to integrate complex layers of knowledge and to develop more refined understanding.

Mathematical or computational frameworks and tools are indispensable in the study of cell regulatory networks [34, 236] because functions, traits and pathologies are rarely caused by single genes [34, 113, 227]. The principal challenge that prevents comprehensive understanding (and simulation) of regulatory networks is their complexity [160]. In the era of systems biology, the icon for molecular biology is the 'hairball' graph, which illustrates how everything seems to interact with almost everything else [69, 146]. High-throughput technologies generate such large volumes of data that there is concern about how to grasp the big picture [37, 62, 121] and most data sets are not being used to their full potential.

We seek better insight into the complexity inherent in non-equilibrium ES cell regulatory networks undergoing lineage specification by developing computer simulations of self-organisation using the CoSMoS approach. Simulation, together with the hypothesis that lineage computation occurs at the edge of chaos, should allow us to investigate the driving of gradual accumulation of network complexity 'from the bottom up'. Here, we present the first steps in this design process: use of the CoSMoS approach to develop a highly abstracted model and simulation of regulatory network activity driven by just pluripotent transcription factors (TFs), at genome-wide scales.

We present two increments of this novel computational framework to interrogate the complexity of stem cell regulatory networks. We employ a

© Springer Nature Switzerland AG 2018
S. Stepney, F.A.C. Polack, *Engineering Simulations as Scientific Instruments: A Pattern Language*, https://doi.org/10.1007/978-3-030-01938-9_11

previously-described theoretical framework based on the notion that the backbone of stem cell fate computation is provided by the critical-like self-organisation of transcription factor (TF) regulatory networks [108–110].

The structure of this part follows the CoSMoS patterns as defined in the CoSMoS approach. Each chapter documents a model and simulator increment, in terms of the three phases of discovery, development, and exploration.

In the first increment (Chapters 12–15), we develop a simulation of a single TF in isolation, and instantiate it with three separate TFs: Oct4, Nanog and Sox2, central elements of the core pluripotent network of mouse embryonic stem cells. The version presented here is a minimally updated version of [98], changed to correct small errors, and to bring in line with the latest version of CoSMoS patterns. In the second increment (Chapters 16–19), we allow instantiations of the three TFs to couple, so that activity of one TF can potentially ignite activity in another. The version presented here is an updated version of [99], changed to bring in line with the latest version of CoSMoS patterns, and to update some of the figures and domain behaviours in line with those presented in [100].

We conclude (Chapter 20) with some reflections on the process, discussion of further work, and information about obtaining and running the simulation code.

Chapter 12
CellBranch: increment 1: single transcription factor

The models and results presented here document the first increment of the CellBranch project using the CoSMoS design cycle. Here, we design and calibrate simulations of single TFs in isolation. This single TF version of the full model is not biologically realistic; its purpose is to serve as a building block of complexity as the basis of following increments.

CoSMoS Simulation Project (92)
Develop a basic fit-for-purpose simulation of the complex scientific domain of interest.

- carry out the Discovery Phase (95)
- carry out the Development Phase (96)
- carry out the Exploration Phase (97)
- Argue Instrument Fit For Purpose (186)

The CellBranch project development omits the CoSMoS argumentation phase.

© Springer Nature Switzerland AG 2018
S. Stepney, F.A.C. Polack, *Engineering Simulations as Scientific Instruments: A Pattern Language*, https://doi.org/10.1007/978-3-030-01938-9_12

Chapter 13
CellBranch: increment 1: Discovery Phase

> **Discovery Phase (95)**
> Decide what scientific instrument to build. Establish the scientific basis of the project: identify the domain of interest, model the domain, and shed light on scientific questions.
>
> - identify the Research Context (119)
> - define the Domain (123)
> - construct the Domain Model (128)

13.1 Discovery Phase > Research Context

> **Research Context (119)**
> Identify the overall scientific context and scope of the simulation-based research being conducted.
>
> - provide a brief *overview* of the research context
> - document the *research goals* and project scope
> - agree the Simulation Purpose (121), including criticality and impact
> - identify the team members and their experience, and assign Roles (99)
> - Document Assumptions (108) relevant to the research context
> - note the available *resources*, timescales, and other constraints
> - design and set up a Project Repository (219)
> - determine *success criteria*
> - decide whether to proceed, or walk away

© Springer Nature Switzerland AG 2018
S. Stepney, F.A.C. Polack, *Engineering Simulations as Scientific Instruments: A Pattern Language*, https://doi.org/10.1007/978-3-030-01938-9_13

13.1.1 Discovery Phase > Research Context > overview

The context of this research is the investigation of a conceptual approach: self-organisation at the edge of chaos. We have argued that if the activity of single transcription factors can be described as critical-like branching processes, their interplay should define a critical-like genome-wide interference pattern that captures in some way the nature of the entire pluripotency transcription factor regulatory network [109].

Here we build a simulation based on the representation of TFs as *branching processes*. The mathematical concept of a branching process (BP) is as follows. Consider a population of individuals. At time t each individual i produces a next generation of m_i offspring individuals, with the value of m_i drawn from some probability distribution. Let the average number of offspring produced be μ. If $\mu > 1$, then the process is supercritical and the number of individuals grows without bound. If $\mu = 1$ then the system is critical and can either give rise to more individuals in the next step or lead to dissipation of the process. If $\mu < 1$ then the process goes to extinction.

Our model of TF BPs builds on this idea, and also allows the TFs to *interact* in such a way as to cause the regulatory network to self-organise at the edge of chaos. We capture the activity of single TFs as BPs in order to predict the interplay of multiple TFs and the emergent nature of the entire TF regulatory network, hypothesised to operate in a critical-like state [109].

For a TF to be stably expressed, its BP must be supercritical [109]. Therefore, by modelling the activity of TFs known to be expressed in mouse embryonic stem cells, we link the perturbation of a TF's cistrome (portion of the genome in which the TF displays some activity) with a dynamic and distributed description of TF activity. This is a prerequisite to being able to simulate the entire TF regulatory network of an ES cell, as argued in [109]. The TFs called Oct4, Sox2 and Nanog are central elements of the core pluripotent network of mouse embryonic stem cells. In this first increment, we develop our simulation for these three TFs in isolation, and so characterise how their associated TF BPs propagate in the absence of interference or communication.

Our incremental approach to the development of the full simulation commences with the simplest possible system: the operation of one transcription factor at genome-wide scales. We later add layers of further complexity, testing and calibrating as we go.

!! Future: A model of a single pluripotent TF in isolation is far from complete and is not biologically realistic. It is only when multiple communicating TF BPs are simulated in parallel that we can expect to generate the interference patterns predicted to underpin circuitry self-organisation. As greater numbers of pluripotency TFs are included in the model, we anticipate that our simulations will become increasingly biologically realistic. In future increments we will augment the complexity of the computational model in a stepwise manner, adding detail and refining assumptions as we progress, and increasingly be able to provide insights not accessible by other means.

13.1.2 Discovery Phase > Research Context > research goals

Computer simulation sidesteps the ethical, moral and political issues surrounding use of human embryos. It therefore represents an alternative route to gaining new insight in to this promising field of regenerative medicine. Our overarching aim is to gain sufficient understanding so that any cell type of therapeutic interest can be generated effectively at will.

The specific research goals of CellBranch are:
1. to create a simulation of Branching Process Theory (BPT) as applied to embryonic stem cell differentiation
2. to use this simulation to validate the application of BPT in this context
3. to make the simulation available for more general use

The specific research goals of this first increment are:
1. to simulate and explore a single TF branching process.

13.1.3 Discovery Phase > Research Context > Simulation Purpose

> **Simulation Purpose (121)**
> Agree the purpose for which the simulation is being built and used, within the Research Context.
>
> - define the *role of the simulation*
> - determine the *criticality of the simulation results*

Role of the simulation

The role of the simulation is exploratory: to provide evidence of the usefulness of BPT as a model of decision making in stem cell differentiation. The simulation will be used to investigate which values of the average branching ratio are required to set up a sustainable TF branching process.

Criticality of the simulation results

The simulation work is being used to explore the suitability of a particular conceptual modelling approach, BPT, in the domain. The simulation results are not safety, security, or financially critical: they will not be used directly in the development of any products.

13.1.4 Discovery Phase > Research Context > Roles

Roles (99)
Assign team members to key roles in the simulation project.

- identify the Domain Scientist (101)
- identify the Domain Modeller (103)
- identify the Simulation Engineer (105)
- identify the Argument Modeller (106)
- identify other *optional roles*
- identify necessary *collaborations* between roles

The main CoSMoS roles are fulfilled by the team members in the following way:
- Domain Scientist (101): Julianne Halley, an expert on BPT as applied to stem cell differentiation, backed up by a domain expert in ES cell biology (Austin Smith), and a data collection expert (Sabine Dietmann)
- Domain Modeller (103): Richard Greaves, with CoSMoS domain modelling experience, backed up by a further CoSMoS modelling expert (Susan Stepney)
- Simulation Engineer (105): Richard Greaves, with agent based simulation engineering experience
- Argument Modeller (106): not used in this project: the lead responsibility to Document Assumptions (108) and to collaborate with relevant Roles to agree justifications and consequences was taken by Stepney

The lead domain scientist (Halley) and the domain modellers (Greaves, Stepney) collaborated closely throughout the development of the domain model, translating and abstracting the conceptual TF BP model into a form suitable for simulation.

The domain scientists (Halley, Smith, Dietmann) collaborated on refining the research context.

The simulation engineer (Greaves) collaborated with the the data collection expert (Dietmann) on the form and content of the biological data provided.

13.1.5 Discovery Phase > Research Context > Document Assumptions

---

Document Assumptions (108)
Ensure assumptions are explicit and justified, and their consequences are understood.

- identify that an assumption has been made, and record it in an appropriate way
- for each assumption, determine its nature and criticality
- for each assumption, document the reason it has been made
- for each reason, document its justification, or flag it as "unjustified" or "unjustifiable"
- for each assumption, document its connotations and consequences
- for each critical assumption, determine the connotations for the scope and fitness-for-purpose of the simulation
- for each critical assumption, achieve consensus on the appropriateness of the assumption, and reflect this in fitness for purpose arguments
- revisit the Research Context (119) in light of the assumption, as appropriate

---

A.1 Cistrome data can be provided by processed ChIP-Seq data

   reason It is the data we have
   justification This is one standard use for ChIP-Seq data
   consequence ChIP-Seq data is variable across measurements, so we will need to check the robustness of our results to this variation

A.2 It is sufficient to consider only the key pluripotency transcription factors: Nanog, Oct4, Sox2

   reason As a first step in providing insight, we consider the three TFs widely acknowledged to be central components of the core pluripotent network
   justification See, for example, [36]

consequence We will not be able to determine the effect of further TFs. However, it should be straightforward to incorporate further TF data into the multi-cistrome model

A.3 We can use mouse data as a suitable proxy for data from human ES cells

reason Suitable mouse data is more readily available; mouse ES cells have an unambiguous 'ground state'; so mouse data is a good basis for evaluating the TF BP model

justification Although effective manipulation of human ES cells is a long term goal, here we are only assessing the TF BP model

consequence We cannot extrapolate results to the human system

13.1.6 Discovery Phase > Research Context > resources, timescales, other constraints

1. The project has a one year duration. The Domain Scientist is employed full time, and Simulation Engineer part time.
2. The work has access to a local computer cluster, for running simulations and gathering performance metrics.
3. The team members are split between York (Halley, Greaves, Stepney) and Cambridge (Smith, Dietmann).

13.1.7 Discovery Phase > Research Context > Project Repository

> **Project Repository (219)** : Use a project-wide repository to coordinate the project information.

We use a shared Google Drive for documentation and notes, and a GitHub repository for version-controlled code.

13.1.8 Discovery Phase > Research Context > success criteria

1. a single-cistrome simulator that exhibits the expected behaviours, and can be used as the basis for multi-cistrome simulator development
2. a single-cistrome simulator that can justify the use of the TF BP model to analyse stem cell fates

13.1.9 Discovery Phase > Research Context > decide

At this point, the decision is made to proceed with the CoSMoS Simulation Project.

13.2 Discovery Phase > Domain

> **Domain (123)**
> Identify the subject of simulation: the real world biological system, and the relevant information known about it.
>
> - draw an explanatory Cartoon (124) of the domain
> - provide an overview description of the domain
> - define the Expected Behaviours (126)
> - provide a Glossary (127) of relevant domain-specific terminology
> - Document Assumptions (108) relevant to the domain
> - define the scope and boundary of the domain – what is inside and what is outside – from the Research Context (119)
> - identify relevant sources: people, literature, data, models, etc.

13.2.1 Discovery Phase > Domain > Cartoon

> **Cartoon (124)** : Sketch an informal overview picture of the Domain.

Figure 13.1 is a cartoon of the regulatory process. A single gene regulation and its expression is conceptually relatively straightforward; the complex interplay of multiple interacting regulatory processes is not.

13.2.2 Discovery Phase > Domain > overview

Embryonic stem (ES) cell biology

Modern, high-throughput laboratory techniques routinely provide large-scale datasets including complete genome sequences, dynamic measurements of gene expression, extensive lists of regulatory proteins and RNAs, and *in vivo* occupancy of DNA by TFs, cofactors and nucleosomes [17]. Such datasets facilitate the investigation of ES cell regulatory networks. To cre-

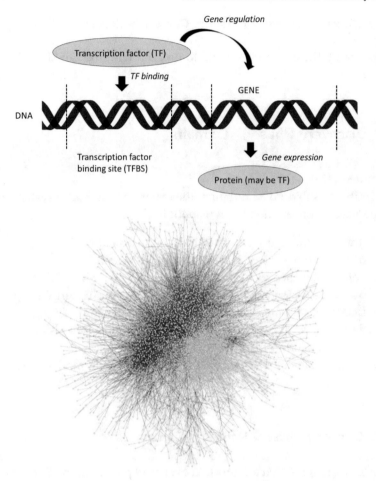

Fig. 13.1 Domain> Cartoon: (top) The regulatory process: a TF protein binds to DNA at the binding site, thereby regulating production of protein (which may be a TF) from the corresponding gene (gene expression). (bottom) A 'hairball graph' of the human proteome and its binding interactions; expressed proteins may include other TFs that can regulate expression of other genes [69, Fig. 1]

ate a complete multi-layered model of a stem cell regulatory network one should exploit these big data to bridge gaps between the phenotypic behaviour of whole cells and key regulatory molecules [235].

We need to capture the results of multiple high-throughput experiments within a logical and transparent conceptual and computational framework in order to facilitate the interrogation of multiple layers of complex regulatory information. Our initial model is based on the complete genome sequence of mouse embryonic stem cells and on ChIP-Seq data that capture the density of TF binding sites throughout the genome. TFs operate in parallel, influencing each other; according to our hypothesis, they produce genome-wide interference patterns that capture in some way the hypothesised nature of the entire pluripotent circuitry.

Embryonic stem (ES) cells have the potential to produce all of the different cell types within the body, but this behaviour cannot yet be efficiently exploited *in vitro*. We have considerable knowledge of the component parts of the regulation of ES cells maintained under precise external conditions [155], but during normal development many different types of regulatory factors interact, enabling cells to respond flexibly to changing environments. The regulatory network of single ES cells is therefore some function of both cell intrinsic and cell extrinsic variables.

Here we assume that pluripotency is a state of individual ES cells. ES cells exit pluripotency via a transient 'primed' state that facilitates cell fate computation [166]. Our knowledge of this exit process and the transient primed state is incomplete, partly because it is difficult to obtain data from transient cell states [215]. The process of pluripotency exit itself is intrinsically disorganised and/or chaotic in order for it to integrate intrinsic and extrinsic information and compute cell fate. According to our conceptual framework, regulatory circuitries compute cell fate trajectories via 'critical-like dynamics' at the edge of chaos [109].

Nanog, Oct4 and Sox2 form part of the core pluripotency circuitry of ES cells [36]. Oct4 in particular seems central to understanding pluripotency. Oct4 expression level is closely regulated, with deviations either above or below a certain expression range resulting in differentiation [170]. It has been suggested that protein complexes, in which Oct4 is involved, help to establish a dynamic competition between individual elements, serving to buffer the differentiation-promoting activity of Oct4 [164].

Fluctuations are inevitable in any system that has many degrees of freedom. At static equilibrium, such fluctuations ultimately disappear but under non-equilibrium conditions, fluctuations are often great enough to drive reorganisation toward new dynamic states [48, 168]. If continual driving is experienced, complex spatiotemporal patterning usually results and systems are said to have 'self-organised' [21, 94, 168].

In biology, the growth and development of organisms occurs far from equilibrium. The stem cell regulatory networks that facilitate these processes are replete with positive and negative feedback loops and nonlinear interactions. When faced with overwhelming complexity, the natural tendency of humans is to either reduce, simplify or ignore it. Reductionist thinking makes systems (a) easier to think about, (b) easier to consider manipulating, and (c) easier to predict, provided non-equilibrium driving is minimal.

Over the last few decades, there has been increasing awareness of the limitations of the reductionist approach [18, 57, 68, 134, 173] and it has become clear that some laws of nature cannot be deduced by resolving more detail [222]. This so called 'new era of physics' focuses on developing complex behaviour out of simplicity, instead of the traditional reductionist approach that reduced complexity to its simplest possible form [9, 130, 173]. Non-equilibrium driving can have profound consequences on system behaviour, a realisation that contrasts with our natural tendency to assume systems are near equilibrium or at least show some steady state behaviour. Equilibrium and reductionist thinking pervades most scientific disciplines [19–21, 64], including molecular and stem cell biology.

The differentiation of pluripotent cells in the early embryo is a fascinating non-equilibrium process that results in the production of numerous specialised cell types. More than 600 different proteins have been implicated in exit from a naïve pluripotent state and control of early state transitions in the mouse [131]. As our focus shifts from individual components to complex communication networks, experimental studies have become more difficult. Not only do central features of complex networks, such as robustness, prevent straight forward analysis and interpretation of network behaviours, but many experiments cannot be performed because of ethical reasons surrounding the use of human embryos. Computer simulation sidesteps these ethical, moral and political issues.

13.2.3 Discovery Phase > Domain > Expected Behaviours

> **Expected Behaviours (126)** : Describe the hypothesised behaviours and mechanisms.

In the single cistrome case, we expect low values of TF expression to be insufficient to sustain activity, and higher values to be sufficient to sustain activity.

13.2.4 Discovery Phase > Domain > Glossary

> **Glossary (127)** : Provide a common terminology across the simulation project.

The main biological terms used in the various models are:

binding site : section of DNA that binds a given TF and influences transcription of associated genes

branching process (BP) : the mathematical model underlying inspiration of the TF BP framework being investigated here

ChIP-Seq : a technique to identify the binding sites of transcription factors on DNA

cistrome : the portion of the genome associated with a specific TF; a pattern of genome-wide binding sites to which the TF displays some activity

pluripotent stem cell : a cell capable of generating all the cell types present in the adult body

segment : the genome data is segmented, into say 10k or 50k base-pair sequences, in order to apply the TFBP framework

transcription factor (TF) : a protein that binds to DNA to influence transcription of the associated gene

13.2.5 Discovery Phase > Domain > Document Assumptions

See §13.1.5 for the Document Assumptions (108) pattern requirements.

A.4 The genome can be modelled as a set of overlapping TF cistromes without needing epigenetic factors

> **reason** We are looking only at TF segments, and the pluripotent state can be induced by TFs alone
> **justification** See, for example, [138]
> **consequence** Behaviours facilitated by other factors, such as epigenetics, will be unseen in the model

A.5 a TF binding site is either bound or unbound, there is no partial TF binding

> **reason** not enough data to say otherwise

A.6 a segment can be either active or inactive; there are no differing amounts of activation

> **reason** Simplification: the data does say whether a segment has one or more binding sites
> **justification** This is the first increment; we may revisit the necessity and impact of this assumption in later increments
> **consequence** We will not be able to separate out behaviours of groups of genes in a segment. In order to do so, we could use smaller segments. But segments cannot be made too small, else we would lose correlations between related TFs.

A.7 we can investigate cell decision making by modelling an individual cell, not a population

> **reason** cells have internal decision making, although they can also be influenced by their environment
> **justification** See, for example, [148]
> **consequence** We will not be able to investigate population-level decision making

13.2.6 Discovery Phase > Domain > scope

- single cell model
- single transcription factor model
- **!! Future:** later increments will add more, coupled TFs, and more interacting cells

13.2.7 Discovery Phase > Domain > sources

- Domain Scientists
- Biological literature, as referenced in the various overviews
- ChIP-Seq data for various cistromes (source: Dietmann)

13.3 Discovery Phase > Domain Model

> **Domain Model (128)**
> Produce an explicit description of the relevant domain concepts.
>
> - draw an explanatory Cartoon (124)
> - discuss and choose the domain Modelling Approach (111) and level of abstraction
> - define the Domain Behaviours (135)
> - build the Basic Domain Model (130) using the chosen modelling approach
> - build the Domain Experiment Model (135)
> - build the Data Dictionary (132)
> - build the domain Stochasticity Model (154)
> - Document Assumptions (108) relevant to the domain model

13.3.1 Discovery Phase > Domain Model > Cartoon

See §13.2.1 for the Cartoon (124) pattern.

Due to the structure of our Basic Domain Model description, the *Domain Model Cartoon* is presented in the section on the TF BP model (Figure 13.4), and should be read in that context.

13.3.2 Discovery Phase > Domain Model > Modelling Approach

> **Modelling Approach (111)** : Choose an appropriate modelling approach and notation.

A central part of this design process is to develop the simplest possible working model at each stage of the modelling process. This 'agile' approach ensures that simulation code is not unnecessarily complicated. It also helps to ensure that if a coding problem is found, it is simple matter to backtrack to the last working model.

The domain model is captured using UML, in anticipation of an agent-based, object-oriented design and implementation of the simulator.

13.3.3 Discovery Phase > Domain Model > Basic Domain Model

> **Basic Domain Model (130)** : Build a detailed model of the basic low level domain concepts, components and processes.

Our domain modelling gives rise to several models at different levels of abstraction: a specifically biological stem cell model of regulatory networks, a model simplifying detailed transcription regulatory networks using branching process theory, and a generic abstract model, which we refer to as the 'sparking posts' model.

Note that the sparking posts model could also be used as a domain model for other biological phenomena as captured by branching process theory, such as patterns of information flow in the human brain.

Regulatory network

We have mouse genome data including the suite of binding sites within it. For convenience and simplicity, we divide this sequence in to 50 kilobase (kb) segments, any of which may or may not contain binding sites for a

■   no TF binding sites

☐   1 or more TF binding sites

▧   target involved in regulation of transcription

Fig. 13.2 A representation of a set of ChIP-Seq data for a cistrome (part of the genome relevant to a specific TF). Each square represents a 50kb segment of DNA. A white square is a segment that contains at least one binding site for product that is not a TF. A red square is a segment that contains at least one binding site for a product that is a TF. A black square is a segment that does not belong to this cistrome

particular TF of interest. If a 50kb segment contains a binding site for our transcription factor, X, then the segment is said to be part of the X cistrome.

Data about the locations of the transcription factor binding sites, in relation to the gene segments in the model, is provided experimentally by ChIP-Seq data. Figure 13.2 is a representation of ChIP-Seq data.

The regulatory network components can be captured in a model such as that shown in Figure 13.3. However, we abstract away from many of these 'hairball' inducing details, and consider the system instead in terms of the TF BP model.

Transcription Factor Branching Process (TF BP) model

A common approach to understanding cell regulatory processes is the application of concepts, tools and techniques developed in mathematics, physics or computer science [149]. Network representations, for example, can ac-

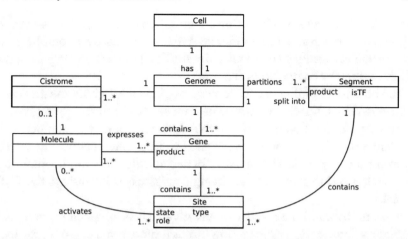

Fig. 13.3 Stem cell pluripotency regulatory network model: class diagram. The stem cell has a genome comprised of genes, which can alternatively be described as a cistrome (or set of cistromes), each being comprised of segments of gene which may or may not contain transcription factor binding sites

commodate multiple types of data within a single visual illustration that provides an overview of regulatory pathways and components [82, 149]. Empirically-derived interaction networks can be difficult to interpret, often appearing as a 'hairball' graph as regulatory mechanisms are increasingly dissected.

We use here a novel way to visualise and simulate genome-wide regulatory network interactions. Our coarse-grained approach does not require details of binding constants prerequisite for most ODE models of stem cell regulation. In many previous computational or mathematical models of stem cell regulatory networks, TFs are represented as single nodes with binary (*on/off*) behaviour. Here, we use a different approach that captures TF activity as a dissipative branching process that propagates within the bounds imposed by the TF's unique cistrome.

Unlike reductionist models that capture TF activity using single variables in an equation, in our model we explicitly represent a background delocalisation of TF activity throughout the genome. We can visualise the activity of each TF's BP as a kind of gateway through which regulatory information pertaining to the TF passes over time.

The TF BP model allows a decoupling between details of binding site constants and the emergent effect of TF activity throughout the genome. Instead of struggling with countless (often unknown) binding constants, we consider the overall flow of regulatory information at genome-wide scales. It is thus more suitable for attempts to discover how the ES cell regulatory network behaves as a whole during computation of lineage choice. Through this more coarse-grained methodology, we hope to discover complex interactions that can easily be overlooked by studies that focus on only a handful of key regulatory components at a time.

The potential binding of a TF to target regions throughout the genome is determined by ChIP-Sequencing. The data set or 'footprint' for a given TF comprises a unique pattern of TF-DNA interactions that is somewhat dependent upon the precise methods used to infer interactions. The precise footprint for a specific TF may vary between different experimental datasets. Such 'fuzziness', rather than being a nuisance, is intrinsic to the TF BP model.

If we understand the activity of any given TF as a branching process of regulatory information propagating through time, it makes sense for there to be some correlation between observed TF expression and the saturation of target sites influenced by TF activity. The significance of this point should become clearer in later increments, when we simulate multiple cistrome data sets. In this increment, we focus on simulating a single TF's BP to introduce the groundwork for our approach.

Figure 13.4 presents a Cartoon of the TF BP model. Each square in the figure corresponds to a 50kb segment of the mouse genome. Black squares represent segments that contain no binding sites for the TF of interest, while red and white squares represent segments with at least one binding site for the TF of interest. The difference between a red and white segment lies in their products. A red segment has products that include TFs, whereas none of the products of a white segment is a TF. Henceforth, when we refer to a 'red segment' we mean a gene segment that can bind TF and thus become stimulated into transcribing further TFs.

We capture the countless (ill-defined or unknown) cascades of gene activation via TF production and feedback as a branching process in which TFs produce other TFs while also regulating the remainder of the genome. There are potentially three qualitatively different types of behaviour for any TF_X branching process. Firstly, the cistrome X is saturated and the TF_X gene is continually and stably expressed. Alternatively, there is the opposite type of emergent behaviour, with TF_X expression occurring at a very low noisy level that is not sustainable unless TF_X is supported by continual activation of the TF_X gene via some external signal. Finally there is a dynamic intermediate between these extremes where a branching process only just percolates through the TF_X cistrome. In all cases, the targets of TF_X are divided in to two types: (1) dissipative targets that do not propagate information back in to the TF_X cistrome, and (2) amplifying targets that are either TFs

time t

time t+1

time t+2

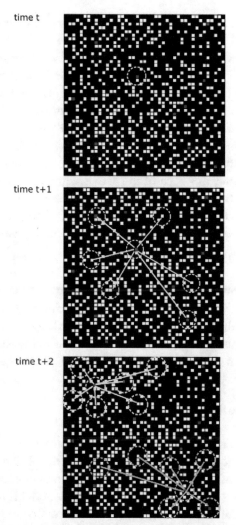

Fig. 13.4 Domain Model > Cartoon: A branching process representation of the overall flow of regulatory information, which serves as the basis of our simulation. At t, assume the circled red segment is active. At time $t + 1$ this activates m further randomly chosen segments (arrows), and itself deactivates. At time $t + 2$, all of these newly active segments that are themselves red each activate a further m randomly chosen segments, and deactivate

themselves and capable of propagating information or code for signalling molecules that are involved in signal transduction.

We define an average branching ratio, called m, for our gene regulation branching process. That is to say that once transcribed, a gene (or gene segment in our case) will produce m product molecules (in this single cistrome model these will all be the TF that binds to binding sites within the cistrome of interest). If the active site is associated with TF products then new TFs are produced and these can bind to other TF binding sites in the system. In this way, up to m segments are activated in the next time step of the algorithm. In the time step after this each of the active segments can go on to activate m further segments and so on as illustrated in Figure 13.4.

This TF BP model is built on the classical BP theory outlined in section 'Domain > overview', and is adapted in the following ways:

- m is related to the BP branching factor μ, but is not the same, because here the m 'offspring' include both white and red segments, yet only red segments go on to produce further 'offspring'.
- In the supercritical case, the number of offspring cannot increase without bound, but only up to the number of relevant segments in the cistrome.
- The individuals are segments, and do not 'die' at the end of a generation; rather they can be reused (reselected) in subsequent generations.

Basic Domain Model: Sparking Posts

In order to model a branching process, we produce our domain model in terms of a metaphor. To capture the nature of critical-like self-organisation hypothesised to underpin lineage computation, we have reduced the system to a 'sparking posts model'. This computational model is used to define the backbone of critical-like self-organisation upon which other layers of complexity are elaborated.

The TF BP representation of our system is modelled as a 'sparking posts' representation of the cistrome in which each segment is modelled as a metal 'post' which emits 'sparks' once it has been activated by an incoming spark emitted by another post in the previous timestep. The sparks represent the TF products of the genes contained within a given segment and are therefore the principal mode of communication between cistromes, the genome being effectively the sum of all cistromes in the system.

So the Basic Domain Model is as follows.

- An *arena* contains metal *posts*, some *red*, some *white*. The arena is an abstraction of a particular cistrome; the posts are abstractions of the segments containing binding sites (red and white squares in Figure 13.2);

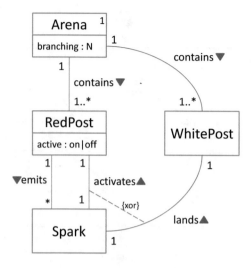

Fig. 13.5 Sparking posts model for a single arena: class diagram. There is one arena, which has a branching factor. The arena contains multiple red posts, which can be on or off, and multiple white posts. A red post can emit several sparks; each spark is emitted by a particular post. A particular spark either activates a red post or lands on white post, but not both

red posts are abstractions of segments that express TFs (red squares in Figure 13.2).

- Posts may be *active* (on) or *inactive* (off). In a timestep, an active red post emits *m sparks*. A post being active is an abstraction of a gene in a segment being active; a red post sparking is an abstraction of an active gene expressing a TF.
- Posts become inactive after they have sparked. A spark lands on a random post in the arena (that is, the model is aspatial), and activates it.
- Continued propagation of sparks relies on the activation of sufficient red posts at each timestep.

Figures 13.5 and 13.6 capture aspects of this Domain Model.

13.3.4 Discovery Phase > Domain Model > Data Dictionary

Data Dictionary (132) : Define the modelling data used to build the simulation, and the experimental data that is produced by domain experiments and the corresponding simulation experiments.

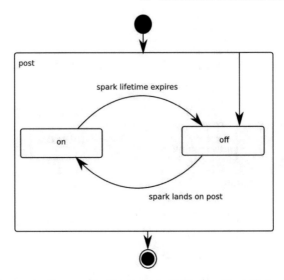

Fig. 13.6 RedPost state diagram. RedPosts are initially inactive (off); become active (on) if a spark lands; then become inactive in the next timestep

p	total number of posts in the arena
r	number of red posts
m	sparks emitted per active red post
s_0	number of red posts active initially
t	timestep
s_t	number of red posts active at timestep t

Fig. 13.7 Data Dictionary: (top) parameters, constant during a simulation run; (bottom) variables, changing during a simulation run

The sparking post model's parameters and variables are given in Figure 13.7. Figure 13.8 gives the values of some of these parameters for the cistromes of interest here.

13.3.5 Discovery Phase > Domain Model > Domain Experiment Model

Domain Experiment Model (135) : Define relevant experiments in the Domain, as the basis for analogous Simulation Experiments and results analyses.

The Domain Model is sufficiently abstracted from the Domain that the Simulation Experiments in this increment do not mirror any domain experiments.

	Nanog	Sox2	Oct4
p	4310	3330	2540
r	631	542	466
r/p	0.146	0.163	0.183
$p/r = m_c$	6.8	6.1	5.5

Fig. 13.8 Data Dictionary: parameter values for p (number of posts, or segments in the cistrome); r (the number of red posts, or red segments in the cistrome); derived value r/p, the proportion of posts that are red; derived value m_c, the critical branching factor in the infinite arena limit

Hence it is unnecessary to build a Domain Experiment Model in this increment.

13.3.6 Discovery Phase > Domain Model > Domain Behaviours

Domain Behaviours (135) : Describe the observed emergent behaviours of the underlying system.

The 'sparking posts' domain model forms the basis for subsequent simulation development.

We can form a much simpler version of the model, which will help to understand the effect of noise. Since there are a finite number of posts, stochastic fluctuations will occur, and sparks might occasionally miss many or all of the red posts. Here we instead assume that posts are always hit the average number of times. This is the case when $p \to \infty$ whilst keeping r/p constant. We are interested in the proportion of red posts active in the 'steady state', in limit of large time.

At time t there are s_t red posts active. Each of these active post emits m sparks, so a total of $s_t \times m$ sparks are emitted. Let each of these sparks be absorbed by a separate post, of which a fraction r/p are red. So at the next timestep, there are $s_{t+1} = s_t mr/p$ red posts active.

The number of active red posts reduces with time if $m < p/r$, and so the arena is extinguished, with $s_\infty = 0$.

The number of active red posts steadily grows with time if $p/r < m$, until there are more sparks emitted than there are posts in total (moving outside our assumption of each spark being absorbed by a separate post), and so the arena saturates with $s_\infty = r$.

The critical value, m_c, where this change of behaviour happens is $m_c = p/r$. Values for m_c for the TFs of interest are shown in Figure 13.8.

Hence the expected behaviour of the single cistrome simulation is to quench for low values of m, saturate for high values of m, and have a tipping point around m_c.

13.3.7 Discovery Phase > Domain Model > Document Assumptions

See §13.1.5 for the Document Assumptions (108) pattern requirements.

First, we have some assumptions related to the TF BP model, which we note as they have an impact on the sparking posts model.

A.8 the product of a TF producing segment is the TF whose cistrome we are modelling

> **reason** An assumption underlying use of the TF BP model
>
> **justification** The TF may not be directly produced; there may be a cascade of production, but the TF BP model collapses this cascade. We are investigating this model
>
> **consequence** This is an abstraction from the biology, made to allow us to model the highly complex processes. If it works, this abstraction could also provide an approach to include other features such as epigenetics and mRNAs in a tractable model

A.9 the identity of the TFs produced during transcription is irrelevant in the single cistrome model

> **reason** An assumption underlying use of the TF BP model
>
> **justification** The TF BP model assumes that the relevant scale of computation is the cistrome level, abstracted from specific details of the individual TFs

Assumptions directly related to the sparking posts model are:

A.10 a spark from a post can hit any post with equal probability: there is no notion of a 'distance' between posts

> **reason** an aspatial model
>
> **justification** the TF BP model collapses a potential cascade of TFs into a single 'proxy' TF. This cascade would lose any spatial dependence in the DNA.

A.11 a post cannot be hit by more than one spark per timestep: there is no notion of different 'capacity' posts

> **reason** follows from assumption A.6

Chapter 14
CellBranch: increment 1: Development Phase

> **Development Phase (96)**
> Build the scientific instrument: produce a simulation platform to perform repeated simulation, based on the output of the Discovery Phase (95).
>
> - revisit the Research Context (119)
> - develop a Platform Model (149)
> - develop a Simulation Platform (161)

14.1 Development Phase > revisit Research Context

The research context is unchanged in the light of *Discovery* phase activities. The TF concepts need to be reinterpreted in terms of the sparking posts model.

14.2 Development Phase > Platform Model

> **Platform Model (149)**
> From the Domain Model, develop a platform model suitable to form the requirements specification for the Simulation Platform.
>
> - choose a Modelling Approach (111) and application architecture for the platform modelling

© Springer Nature Switzerland AG 2018
S. Stepney, F.A.C. Polack, *Engineering Simulations as Scientific Instruments: A Pattern Language*, https://doi.org/10.1007/978-3-030-01938-9_14

- develop the platform model from the Domain Model (128). In particular:

  – remove the Domain Behaviours (135)
  – develop the Basic Platform Model (151) from the Basic Domain Model (130)
  – develop the Simulation Experiment Model (152) from the Domain Experiment Model (135)

- Document Assumptions (108) relevant to the platform model
- if necessary, Propagate Changes (168)

14.2.1 Development Phase > Platform Model > Modelling Approach

Modelling Approach (111) : Choose an appropriate modelling approach and notation.

We use the same approach as for Domain Model (see §13.3.2), assisting Seamless Development.

14.2.2 Development Phase > Platform Model > Basic Platform Model

Basic Platform Model (151)
Build a detailed model of the basic platform concepts, components and processes.

- develop the Basic Platform Model (151) from the Basic Domain Model (130)
- as needed, develop the Stochasticity Model (154)
- as needed, develop the Space Model (155)
- as needed, develop the Time Model (158)

The 'sparking posts' model carries over from the Basic Domain Model unchanged. The emergent tipping point behaviour is not an explicit component of the Basic Platform Model.

Stochasticity Model (154) : Each spark can land on any post with equal (uniform) probability (see assumption A.10).

Space Model (155) : The model is aspatial: although the post metaphor might sound spatial, the equal probability of spark landing implies an aspatial arena.

Time Model (158) : The model of time is that of synchronised timesteps of
activity: posts emit sparks, posts are activated by sparks landing on
them, in each timestep (a consequence of the Basic Domain Model (130)).

14.2.3 Development Phase > Platform Model > Simulation Experiment Model

> **Simulation Experiment Model (152)** : Define relevant experiments in
> the simulation, analogous to domain experiments.

In this case we do not have a relevant Domain Experiment Model to use as
the basis for design. The kinds of Simulation Experiments we will do require
the following input/output and instrumentation:
- derive parameters p, r from ChIP-Seq data
- input and set parameters p, r, m, s_0
- run the simulation for T timesteps
- output s_T, the number of active posts at time T
- perform multiple runs with different random seeds

14.2.4 Development Phase > Platform Model > Document Assumptions

See §13.1.5 for the Document Assumptions (108) pattern requirements.

A.12 the sparks emitted by an active post last for one simulation time step

> **reason** simplicity
> **justification** first increment
> **consequence** half lives and decay rates are not modelled; they may be
> added in later increments

14.3 Development Phase > Simulation Platform

> **Simulation Platform (161)**
> Develop the executable simulation platform that can be used to run the
> Simulation Experiment.

- choose an Implementation Approach (160) for the platform modelling, following the principle of Seamless Development (214) as much as possible
- coding
- testing
- perform Calibration (163)
- Document Assumptions (108) relevant to the simulation platform
- if necessary, Propagate Changes (168)

14.3.1 Development Phase > Simulation Platform > Implementation Approach

Implementation Approach (160) : Choose an appropriate implementation approach and language.

The simulation is implemented as an object-oriented Java application using the MASON simulation environment to handle such things as time-stepping the simulation and on screen graphics (when running in graphical mode).

14.3.2 Development Phase > Simulation Platform > coding, testing

The details are omitted here. The code is available on GitHub for inspection (see §20.2).

14.3.3 Development Phase > Simulation Platform > Calibration

Calibration (163) : Tune the Simulation Platform so that simulation results match the calibration data provided in the Data Dictionary.

The Domain Model is sufficiently abstracted from the Domain that the Simulation Experiments in this increment do not mirror any domain experiments. Hence there is no calibration data in the Data Dictionary, and no Calibration activity.

14.3.4 Development Phase > Simulation Platform > Document Assumptions

There are no further relevant assumptions made in the simulation platform development.

Chapter 15
CellBranch: increment 1: Exploration Phase

Exploration Phase (97)
Use the Simulation Platform resulting from the Development Phase to explore the scientific questions established during the Discovery Phase.

- initially, revisit the Research Context (119)
- develop an experimental Results Model (174)
- finally, revisit the Simulation Purpose (121)

15.1 Exploration Phase > revisit Research Context

The Research Context (119) is unchanged in the light of the Discovery Phase and Development Phase activities.

15.2 Exploration Phase > Results Model

Results Model (174)
Develop a results model suitable for interpreting Simulation Experiment data in Domain Model terms.

- perform Sensitivity Analysis (175)
- perform relevant Simulation Experiment (177)s
- build a Simulation Behaviours (179) model

© Springer Nature Switzerland AG 2018
S. Stepney, F.A.C. Polack, *Engineering Simulations as Scientific
Instruments: A Pattern Language*, https://doi.org/10.1007/978-3-030-01938-9_15

15.2.1 Exploration Phase > Results Model > Sensitivity Analysis

> **Sensitivity Analysis (175)** : Determine how sensitively the simulation
> output values depend on the input and modelling parameter values.

No explicit sensitivity analysis is performed at this stage. Some of the
specific Simulation Experiments below investigate the sensitivity to various
parameters.

15.2.2 Exploration Phase > Results Model > Simulation Experiment

> **Simulation Experiment (177)**
> Design, run, and analyse simulation experiments.
>
> - design the experiment
> - perform simulation runs and gather data
> - analyse results, for input to the Simulation Behaviours (179) model
> - Document Assumptions (108) relevant to the simulation experiment

15.2.3 Exploration Phase > Results Model > Simulation Experiment > design

The parameters p (number of posts) and r (number of red posts) are effectively fixed for any given set of experimentally derived cistrome data (Figure 13.8). We can also generate synthetic data to create systems with a range
of p and r values to explore general behaviours.

Number of simulation runs

We are not performing any statistical analyses at this stage of the project, merely inspecting behaviour. However, the simulation is essentially
stochastic, and when we do come to perform statistics, we will need to
choose the number of runs based on the significance, power, and effect size
of interest. For consistency, we make that choice now, and use the relevant
number of runs.

We require a statistical significance of 99% (a 1% false positive rate), a
statistical power of 99% (a 1% false negative rate), and a 'medium' effect
size (Cohen's $d = 0.5$, the ability to distinguish a difference in means of

0.5 of a standard deviation). Calculating the required sample size for these experimental parameters[i] gives 192.

We round this up, and take the number of runs to be $N = 200$.

15.2.4 Exploration Phase > Results Model > Simulation Experiment > perform

Protocol

One simulation run comprises the p and r values of a particular arena (chosen for example to match Nanog, Sox2, Oct4 data), an m value (0–50), and a starting activity s_0, as detailed in the experiment design tables.

For each simulation run, we record the proportion of active red posts at the final timestep, $T = 1000$.

For each parameter set (p, r, m, s_0), we run the simulation $N = 200$ times. We identify 4 experiments to perform on the single-arena simulation:

Experiment E.0

Effect of m. For each cistrome, create an arena with the relevant p and r values, and $s_0 = r$. Explore the effect of m by locating those values of m for which the system remains fully saturated: all red posts are active at all time steps. Compare this with the expected m_c value (Figure 13.8) for a noiseless system.

E.0	Nanog	Sox2	Oct4
p	4310	3330	2540
r	631	542	466
m	0–50	0–50	0–50
s_0	r	r	r

Experiment E.1

Effect of s_0, sensitivity to initial conditions. Repeat E.0 with a smaller value of s_0.

---

[i] using, for example, the calculator at powerandsamplesize.com/Calculators/ Compare-2-Means/2-Sample-Equality

E.1	Nanog	Sox2	Oct4
p	4310	3330	2540
r	631	542	466
m	0–50	0–50	0–50
s_0	$r/2$	$r/2$	$r/2$

Experiment E.2

Effect of r. Create an arena with the Nanog p value, and a range of r values. At each value of r, determine the values of m for which the system remains saturated throughout the simulation.

E.2	Nanog
p	4310
r	200, 400, 600, 800
m	0–50
s_0	r

Experiment E.3

Effect of noise. Keeping the value of p/r fixed at the Nanog value of $4310/631$, investigate the effect of reducing p. This gives some insight into whether we can use smaller arenas in experiments to improve simulation performance, without affecting the results.

E.3	Nanog	Nanog
p	2000	1000
r	293	146
m	0–50	0–50
s_0	r	r

15.2.5 Exploration Phase > Results Model > Simulation Experiment > analyse results

Experiments E.0 and E.1 results

See Figure 15.1 for the results of the simulation runs.

Starting with only half the posts active makes little difference to the results.

Experiment E.2 results

For experiment E.2, we take $p = 4310$ (as in Nanog), vary r, and examine how the value of m_c changes. We use $s_0 = r$ throughout.

See Figures 15.2–15.3 for the results of the simulation runs.

Recall that the theoretical tipping point value is $m_c = p/r$. So as r increases, m_c should decrease. This is observed (Figure 15.3).

The smaller the value of r, the noisier the behaviour (visible as more extended boxplots in Figure 15.2). This demonstrates how stochastic effects are more prominent when there are fewer red posts available.

Experiment E.3 results

For experiment E.3, we take $p/r = 4310/631$ (as in Nanog), and vary p keeping p/r constant (mimicking a different sized arena but with the same density of red posts). We use $s_0 = r$ throughout.

See Figure 15.4 for the results of the simulation runs; compare with Figure 15.1(top) for the 'full' arena.

The systems tip at the same point, but the behaviour gets noisier as p (and hence r) decreases, and stochastic effects become more pronounced.

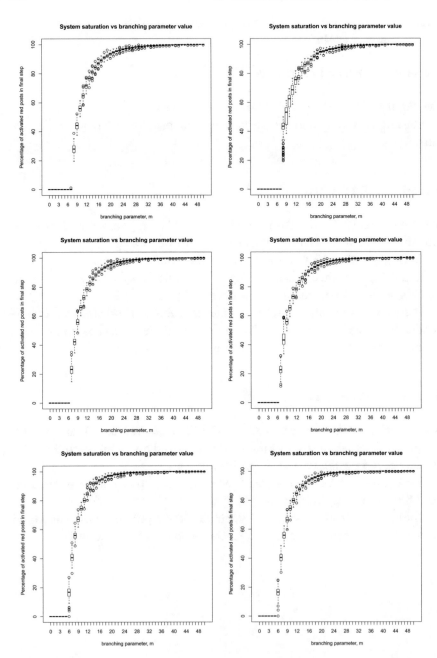

Fig. 15.1 p and r corresponding to (top row) Nanog data, calculated $m_c = 6.8$; (middle row) Sox2 data, calculated $m_c = 6.1$; (bottom row) Oct4 data, calculated $m_c = 5.5$. (left) E.0: $s_0 = r$; (right) E.1: $s_0 = r/2$

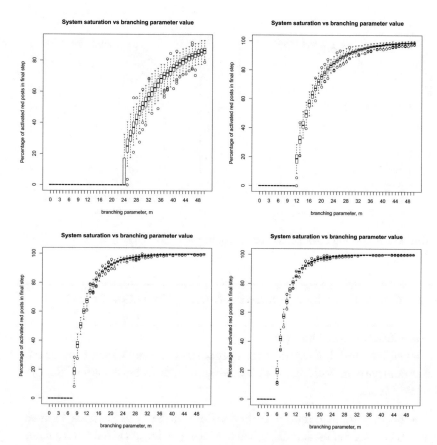

Fig. 15.2 E.2: varying r; here $p = 4310$ (Nanog). (top left) $r = 200$; (top right) $r = 400$; (bottom left) $r = 600$; (bottom right) $r = 800$

r	m obs	m_c calc
200	23–24	21.6
400	11–12	10.8
600	7–8	7.2
800	5–6	5.4

Fig. 15.3 E.2: observed value of m at tipping point for different r (with $p = 4310$, Nanog), versus calculated value m_c

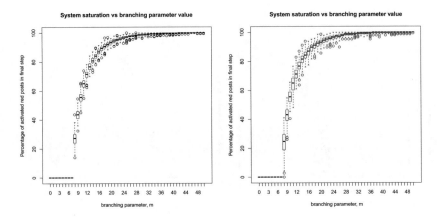

Fig. 15.4 E.3: varying p with constant p/r: (left) $p = 2000, r = 293$; (right) $p = 1000, r = 146$

15.2.6 Exploration Phase > Results Model > Simulation Behaviours

> **Simulation Behaviours (179)**
> Develop a model of the emergent properties of a Simulation Experiment, for comparison with the related emergent Domain Behaviours of the Domain Model.
>
> - build a *minimal* model, from consideration of the Research Context (119), the Simulation Experiment Model (152), the Domain Behaviours (135), and the Calibration (163) translation of the raw simulation data
> - if needed, build an *augmented* model including micro-level observations, and argue the connection to the domain model data
> - if needed, build a Visualisation Model (181)

15.2.7 Exploration Phase > Results Model > Simulation Behaviours > minimal model

The observed values of m where the system 'switches on', and can maintain saturation, are close to the calculated m_c values, see Figure 15.5. However, m has to be somewhat higher than this to saturate the finite-sized arena.

	Nanog	Sox2	Oct4
$p/r = m_c$ calc	6.8	6.1	5.5
m_c observed	8	7	6

Fig. 15.5 Simulation Behaviours: Theoretical value m_c, the critical branching factor in the infinite arena limit, compared to the observed value in the finite-sized arena simulation

15.2.8 Exploration Phase > Results Model > Simulation Behaviours > Visualisation Model

Visualisation Model (181) : Visualise the Simulation Experiment results of the Data Dictionary in a manner relevant to the users.

- The visualisation mimics the cistrome data in Figure 13.2 [plots not shown]
- The output data is presented as plots of activation at $T = 1000$ against m

15.3 Exploration Phase > revisit Simulation Purpose

The purpose of the CellBranch simulation project is to provide evidence of the usefulness of BPT as a model of decision making in stem cell differentiation.

Having exercised the single cistrome simulation, increment 1, we are satisfied that the simulation returns results qualitatively in line with what we expect: simulations run with a higher branching parameter, m, exhibit more sustained activity in that cistrome, and the proportion of the cistrome that contains TF binding site also affects simulation behaviour. Thus we have been able to show that Nanog, Oct4 and Sox2 cistrome branching processes all behave differently, and that each has its own value of m, m_c, at which the simulation runs started to show sustained activity. This observed m_c is marginally higher than the theoretical value, due to noise and finite size effects.

Chapter 16
CellBranch: increment 2: multiple transcription factors

The models and results presented here document the second increment of the CellBranch project using the CoSMoS approach. Here, we increment the model and simulation, by adding multiple interacting TF branching processes, and perform new simulation experiments.

> **CoSMoS Simulation Project (92)**
> Develop a basic fit-for-purpose simulation of the complex scientific domain of interest.
>
> - carry out the Discovery Phase (95)
> - carry out the Development Phase (96)
> - carry out the Exploration Phase (97)
> - Argue Instrument Fit For Purpose (186)

As stated at the start of increment 1, the CellBranch project development omits the CoSMoS argumentation phase.

© Springer Nature Switzerland AG 2018
S. Stepney, F.A.C. Polack, *Engineering Simulations as Scientific Instruments: A Pattern Language*, https://doi.org/10.1007/978-3-030-01938-9_16

Chapter 17
CellBranch: increment 2: Discovery Phase

Discovery Phase (95)

Decide what scientific instrument to build. Establish the scientific basis of the project: identify the domain of interest, model the domain, and shed light on scientific questions.

- identify the Research Context (119)
- define the Domain (123)
- construct the Domain Model (128)

17.1 Discovery Phase > Research Context

Research Context (119)

Identify the overall scientific context and scope of the simulation-based research being conducted.

- provide a brief *overview* of the research context
- document the *research goals* and project scope
- agree the Simulation Purpose (121), including criticality and impact
- identify the team members and their experience, and assign Roles (99)
- Document Assumptions (108) relevant to the research context
- note the available *resources*, timescales, and other constraints
- design and set up a Project Repository (219)
- determine *success criteria*
- decide whether to proceed, or walk away

© Springer Nature Switzerland AG 2018
S. Stepney, F.A.C. Polack, *Engineering Simulations as Scientific Instruments: A Pattern Language*, https://doi.org/10.1007/978-3-030-01938-9_17

The following components of the Research Context are unchanged from increment 1:

- Simulation Purpose (121)
- team Roles (99)
- resources
- Project Repository (219)
- proceed

Components of the Research Context that are changed and expanded for increment 2 are discussed below.

17.1.1 Discovery Phase > Research Context > overview

The overall research context remains much as it was in increment 1: the investigation of the conceptual branching process approach. We have argued that if the activity of single transcription factors can be described as critical-like branching processes, their interplay should define a critical-like genome-wide interference pattern that captures in some way the nature of the entire pluripotency transcription factor regulatory network [109].

We now develop the models and simulation resulting from the first increment, and increase their biological relevance by permitting the system to consist of two or more interacting transcription factor branching processes (TF BPs), thus permitting us to gain understanding of the behaviour of constructively interfering branching processes. We defer inclusion of destructive interference between TF BPs until a later increment, to permit us to more fully understand this simpler representation of the system prior to the addition of another layer of complexity.

This second increment of model development will permit us to characterise the behaviour of the central elements of the core pluripotent network of mouse embryonic stem cells, that is, to characterise the associated TF BPs and how they propagate in the presence of cross-cistrome communication.

Our incremental approach to the development of the full simulation continues with the simplest possible augmentation of the system: the cooperation of two or more transcription factors at genome-wide scales.

This model of multiple interacting pluripotent TF BPs is still far from complete, and not biologically realistic. It is only when multiple TF BPs are simulated in parallel, generating branching process interference patterns via both constructive and *destructive* interference, that we can expect to generate the interference patterns predicted to underpin circuitry self-organisation. This increment allows the simulation of multiple cistromes interacting constructively. As greater numbers of pluripotency TFs are included in the model, we expect that our simulations will become increasingly biologically realistic.

17.1.2 Discovery Phase > Research Context > research goals

Our overall research goals remain unchanged from those stated in increment 1. The specific research goals of this second increment are:

1. to simulate and explore a multi-TF branching process.

17.1.3 Discovery Phase > Research Context > Document Assumptions

> **Document Assumptions (108)** : Ensure assumptions are explicit and justified, and their consequences are understood.

The key assumptions made in the first increment of the simulation development remain relevant, as do their justifications and consequences. In addition:

A2.1 It is sufficient to consider only constructive interference between cistromes

>   **reason** As part of an incremental development of providing insight.
>   **justification** This is a sensible increment that will provide further insight.
>   **consequence** The branching processes can interfere both constructively and destructively [109], so the results of this increment will lack full biological relevance.

17.1.4 Discovery Phase > Research Context > success criteria

1. a multi-cistrome simulator development
2. a multi-cistrome simulator that can justify the use of the TF BP model to analyse stem cell fates

17.2 Discovery Phase > Domain

Domain (123)

Identify the subject of simulation: the real world biological system, and the relevant information known about it.

- draw an explanatory Cartoon (124) of the domain
- provide an overview description of the domain
- define the Expected Behaviours (126)
- provide a Glossary (127) of relevant domain-specific terminology
- Document Assumptions (108) relevant to the domain
- define the scope and boundary of the domain – what is inside and what is outside – from the Research Context (119)
- identify relevant sources: people, literature, data, models, etc.

The following components of the Domain are unchanged from increment 1:

- Cartoon
- Glossary
- Document Assumptions
- sources

Components of the Domain that are changed and expanded for increment 2 are discussed below.

17.2.1 Discovery Phase > Domain > overview

Embryonic stem (ES) cell biology

See the presentation of an overview of the relevant biology in increment 1. We continue to seek to exploit the big data available to understand the phenotypic behaviour of entire cells in terms of the behaviour of key regulatory molecules [235] via creation of a multi-layered model of a stem cell regulatory network.

17.2.2 Discovery Phase > Domain > Expected Behaviours

> **Expected Behaviours (126)** : Describe the hypothesised behaviours and mechanisms.

In the multi-cistrome case, we expect some values of TF expression in one cistrome to be sufficient to sustain activity in another cistrome that has a low level of self-activity. Also, we expect one cistrome to be able to 'ignite' another.

17.2.3 Discovery Phase > Domain > scope

- a single cell model
- multiple TFs interfering constructively
- **!! Future:** later increments may add destructive interference
- **!! Future:** later increments may add more biological detail and build towards a model of interacting cells

17.3 Discovery Phase > Domain Model

> **Domain Model (128)**
> Produce an explicit description of the relevant domain concepts.
>
> - draw an explanatory Cartoon (124)
> - discuss and choose the domain Modelling Approach (111) and level of abstraction
> - define the Domain Behaviours (135)
> - build the Basic Domain Model (130) using the chosen modelling approach
> - build the Domain Experiment Model (135)
> - build the Data Dictionary (132)
> - build the domain Stochasticity Model (154)
> - Document Assumptions (108) relevant to the domain model

The following components of the Domain Model are unchanged from increment 1:
- Cartoon
- Modelling Approach
- Document Assumptions

- (lack of) Domain Experiment Model

Components of the Domain Model that are changed and expanded for increment 2 are discussed below.

17.3.1 Discovery Phase > Domain Model > Basic Domain Model

> **Basic Domain Model (130)** : Build a detailed model of the basic low level domain concepts, components and processes.

As discussed in increment 1, our domain modelling gives rise to several models at different levels of abstraction: a specifically biological stem cell model of regulatory networks, a model simplifying detailed transcription regulatory networks using branching process theory, and a generic abstract model, which we refer to as the 'sparking posts' model. Here we describe the changes to the increment 1 model that arise from allowing multiple constructively interfering cistromes.

Regulatory network

We have mouse genome data including the suite of binding sites within it. For convenience and simplicity, we divide this sequence in to 50 kilobase (kb) segments, any of which may or may not contain binding sites for a particular TF of interest. If a 50kb segment contains a binding site for our transcription factor, X, then the segment is said to be part of the X cistrome.

Data about the locations of the transcription factor binding sites, in relation to the gene segments in the model, is provided experimentally by ChIP-Seq data as in the work described in increment 1. Figure 13.2 is a representation of ChIP-Seq data.

Transcription Factor Branching Process model

We continue to use the Transcription Factor Branching Process model described in increment 1. This novel, coarse-grained approach does not require details of binding constants prerequisite for most ODE models of stem cell regulation. As in the original implementation of the simulation, the refined simulation will also explicitly represent a background delocalisation of TF activity throughout the genome.

Basic Domain Model: Sparking Posts

In increment 1, in order to model a branching process, we produced our domain model in terms of a metaphor.

To capture the nature of critical-like self-organisation hypothesised to underpin lineage computation, we reduced the system to a 'sparking posts' model. This computational model was used to define the backbone of critical-like self-organisation upon which this and other layers of complexity are elaborated.

So, to re-iterate the Domain Model used as the basis for our simulation implementation: The TF BP representation of our system is modelled as a 'sparking posts' representation of the cistrome in which each segment is modelled as a metal 'post' which emits 'sparks' once it has been activated by an incoming spark emitted by another post in the previous timestep. The sparks represent the TF products of the genes contained within a given segment and are therefore the principal mode of communication between cistromes, the genome being effectively the sum of all cistromes in the system.

So the Basic Domain Model with multiple TFs is as follows (see also Figure 17.1).

- An *arena* contains metal *posts*, some *red*, some *white*. There are several arenas; there are some red posts that appear in the same position in different arenas: these are called *shared* posts. An arena is an abstraction of one particular cistrome; the posts are abstractions of the segments containing binding sites (red and white squares in Figure 13.2); red posts are abstractions of segments that express TFs (red squares in Figure 13.2); shared red posts are abstractions of the same segment expressing multiple TFs related to different cistromes.
- Posts may be *active* (on) or *inactive* (off). In a timestep, each active red post emits *m sparks*. A post being active is an abstraction of a gene in a segment being active; a red post sparking is an abstraction of an active gene expressing a TF.
- Posts become inactive after they have sparked.
- The emitted sparks lands on random posts in the arena (that is, the model is aspatial). If a spark lands on an unshared red post, it activates it. If a spark lands on a shared red post, then the spark is transferred to any inactive corresponding post in another arena; if all the shared posts in other arenas are active, the spark activates the post in the original arena. That is, the spark is transferred to another arena where possible. No post can accept more than one spark, whether it shares it or not.
- Continued propagation of sparks in an arena relies on the activation of sufficient red posts at each timestep. Sharing sparks between arenas

1 Suppose we have 3 arenas
that all share a post...

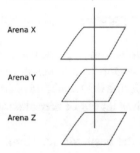

Arena X

Arena Y

Arena Z

2 Suppose that now, a spark
strikes the post in Arena X

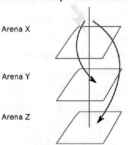

Arena X

Arena Y

Arena Z

Since neither Arena Y nor Arena Z
has a spark at this post, Arena X can
transfer its spark to either Arena Y or
Arena Z

3 Suppose that now, a spark
strikes the post in Arena X
and in Arena Y e.g. when
first 'igniting' the Arenas

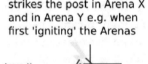

Arena X

Arena Y

Arena Z

Since Arena Z has no spark at
this post, either Arena X or Arena Y
can transfer its spark to Arena Z

4 Suppose that now, sparks
strike the post in all three Arenas
e.g. when first 'igniting' the Arenas

Arena X

Arena Y

Arena Z

Since all the Arenas now have a
spark at this post, none of them
can accept the spark from either of
the others. So all three Arenas retain
the spark...

Fig. 17.1 Domain Model: illustration of communication between arenas: (1) a shared red post: an abstraction of a TF binding site common to two or more cistromes; (2) a shared post is struck by one spark, in arena X; it will activate the post in arena Y or Z; (3) a shared post is struck by two sparks, in arenas X and Y; they will activate the post in arena Z (by transferring the spark from X or Y) and in arena X or Y (the other spark cannot be transferred); (4) a shared post is struck by a spark in each arena, and the post is activated in each arena

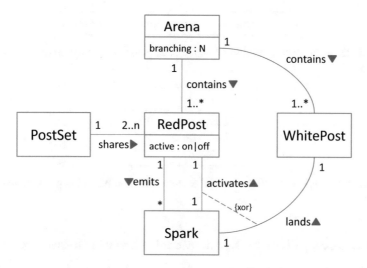

Fig. 17.2 Sparking posts model components for multiple arenas with shared posts: class diagram. Each arena has a branching factor. Each arena contains multiple red posts, which can be on or off, and multiple white posts. A red post can emit several sparks; each spark is emitted by a particular post. A particular spark either activates a red post (Figure 17.1 illustrates which arena this red post is in) or lands on white post, but not both. A shared set of posts comprises two to n red posts, where n is the number of arenas (every red post in a shared set is in a distinct arena; not shown in the diagram); a shared post is in a particular shared set

allows an arena to become or stay active even if it does not produce enough sparks itself.

Figure 17.1 illustrates spark sharing; Figure 17.2 shows the updated class diagram (compare Figure 13.5).

17.3.2 Discovery Phase > Domain Model > Data Dictionary

Data Dictionary (132) : Define the modelling data used to build the simulation, and the experimental data that is produced by domain experiments and the corresponding simulation experiments.

The sparking post model's parameters and variables are as in increment 1, plus the shared posts data (given in Figure 17.3). Figure 17.4 gives the experimentally measured values of σ for the cistromes of interest here.

σ	number of red posts shared by the arenas

Fig. 17.3 Data Dictionary: additional increment 2 parameter, constant during a simulation run

arena	σ
Nanog–Oct4	194
Nanog–Sox2	287
Oct4–Sox2	237

Fig. 17.4 Data Dictionary: Parameter values for σ (the number of red posts shared by the arenas)

17.3.3 Discovery Phase > Domain Model > Domain Behaviours

> **Domain Behaviours (135)** : Describe the observed emergent behaviours of the underlying system.

With multiple interconnected arenas, we expect sparking behaviour in any arena to be affected by the behaviour in other arenas with shared posts.

In increment 1 we use an infinite limit model to calculate an estimate of m_c in a single arena. We here use a similar approach to calculate an estimate in the reduction of m_c in coupled arenas. This infinite limit case is essentially noiseless, with each post being activated the average number of times.

Consider the case of two arenas, X and Y. At time t let there be s_t^X posts active in arena X. In the model, each of these active posts emits m^X sparks, so a total of $st^X \times m^X$ sparks are emitted. Let each of these sparks be absorbed by a separate post with uniform probability. There are three cases (three kinds of posts):

1. Activate a Y arena (red, shared) post. A fraction σ/p^X of the (red) posts are shared, and the model assumes that a spark absorbed by one of these passes to the Y arena. So $st^X m^X \sigma/p^X$ sparks in total are passed to the Y arena, activating that number of posts in the Y arena in the next timestep.
2. Activate an X arena (red, unshared) post. A fraction $(r^X - \sigma)/p^X$ of the (red) posts can be activated in the X arena, and a spark absorbed by such a post activates it in the next timestep. So $st^X m^X (r^X - \sigma)/p^X$ posts in the X arena are activated by these sparks.
3. Absorbed by a white post. The remaining fraction of posts absorb a spark and do not produce a spark in the next timestep.

		arena X	Nanog	Sox2	Oct4
		p/r	6.8	6.1	5.5
arena Y	p/r	m_c	8	7	6
Nanog	6.8	8	–	4.9	4.0
Sox2	6.1	7	6.0	–	4.6
Oct4	5.5	6	6.3	5.6	–

Fig. 17.5 Domain Behaviours: The predicted change in critical branching factor in coupled arenas. Arena Y is held at its observed critical value m_c; the predicted value of cistrome X's m_c when the arenas are coupled is lower than its value when arena X evolves in isolation

Similar arguments, *mutatis mutandis*, hold for sparks emitted by arena Y. So at timestep $t + 1$, the number of active posts in arena X is those activated from arena Y plus those activated from arena X:

$$s^X_{t+1} = \frac{s^X_t m^X (r^X - \sigma)}{p^X} + \frac{st^Y m^Y \sigma}{p^Y}$$

A similar equation holds for s^Y_{t+1}. At the critical tipping point, $s_{t+1} = s_t$. We take these two equations, eliminate s^X/s^Y, then solve for m^X, to get

$$m^X = \frac{p^X \left(p^Y - m^Y r^Y + m^Y \sigma \right)}{\left(p^Y - m^Y r^Y \right) \left(r^X - \sigma \right) + m^Y r^X \sigma}$$

This gives the infinite limit predicted value of m^X in the case of a given m^Y. If we substitute $m^Y = p^Y/r^Y$, the infinite limit single arena critical value for Y, we get $m^X = p^X/r^X$. That is, in the infinite limit, the critical values are unchanged. Alternatively, if we substitute $\sigma = 0$ (isolated arenas), we also recover the original predicted value of m^X. However, if we substitute the *observed* critical value in the finite sized noisy case for m^Y, and the experimentally measured value of σ (shown in Figure 17.4) we get a different prediction for m^X, as shown in Figure 17.5.

17.3.4 Discovery Phase > Domain Model > Document Assumptions

Document Assumptions (108) : Ensure assumptions are explicit and justified, and their consequences are understood.

The assumptions in increment 1 are still relevant to the increment 2 model. We add a further assumption to allow for communication between cistromes in the model:

A2.4 in any given timestep, a post in a cistrome can gain at most one spark from being hit and from sharing sparks from shared posts.

      reason follows from assumption A.6 ('no differing amounts of activation') and assumption A.11 ('a post cannot be hit by more than one spark per timestep')

Chapter 18
CellBranch: increment 2: Development Phase

> **Development Phase (96)**
> Build the scientific instrument: produce a simulation platform to perform repeated simulation, based on the output of the Discovery Phase (95).
>
> - revisit the Research Context (119)
> - develop a Platform Model (149)
> - develop a Simulation Platform (161)

18.1 Development Phase > revisit Research Context

The research context is unchanged in the light of Discovery Phase activities.

18.2 Development Phase > Platform Model

> **Platform Model (149)**
> From the Domain Model, develop a platform model suitable to form the requirements specification for the Simulation Platform.
>
> - choose a Modelling Approach (111) and application architecture for the platform modelling
> - develop the platform model from the Domain Model (128). In particular:

© Springer Nature Switzerland AG 2018
S. Stepney, F.A.C. Polack, *Engineering Simulations as Scientific Instruments: A Pattern Language*, https://doi.org/10.1007/978-3-030-01938-9_18

> – remove the Domain Behaviours (135)
> – develop the Basic Platform Model (151) from the Basic Domain Model (130)
> – develop the Simulation Experiment Model (152) from the Domain Experiment Model (135)
>
> • Document Assumptions (108) relevant to the platform model
> • if necessary, Propagate Changes (168)

The components are realised as follows:
- Modelling Approach: unchanged from increment 1
- Basic Platform Model: The increment 2 'sparking posts' Basic Domain Model, with multiple arenas and shared posts, carries over unchanged
- Simulation Experiment Model: the increment 1 model is unchanged, except for the addition of
    - derive shared r values from ChIP-Seq data, for each arena
    - input and set shared r values for each arena
    - output s_t, the number of active posts at each timestep up to time T
- Document Assumptions: unchanged from increment 1

18.3 Development Phase > Simulation Platform

> **Simulation Platform (161)**
> Develop the executable simulation platform that can be used to run the Simulation Experiment.
>
> • choose an Implementation Approach (160) for the platform modelling, following the principle of Seamless Development (214) as much as possible
> • coding
> • testing
> • perform Calibration (163)
> • Document Assumptions (108) relevant to the simulation platform
> • if necessary, Propagate Changes (168)

The following components of the Simulation Platform are unchanged from increment 1:
- Implementation Approach
- Calibration
- Document Assumptions

Components of the Simulation Platform that are changed and expanded for increment 2 are discussed below.

18.3.1 Development Phase > Simulation Platform > coding, testing

The coding details are omitted here. The code is available on GitHub for inspection (see §20.2).

The behaviour of the increment 2 simulation experiments should reduce to that of increment 1 in the case of a single arena. As part of the testing process, we have re-run some of the increment 1 experiments, requiring the same results.

Testing experiment E.1

We repeat experiment E.1 from increment 1 with each of the three core pluripotency cistromes to verify that the modified simulation returns results consistent with those of the previous increment. As before, these simulations commence with $s_0 = 0.5r$ and m is varied between 0 and 50 in an attempt to locate the critical value of m, m_c, at which we first start to observe sustainable branching in the cistrome of interest.

The results obtained are shown in Figure 18.1, and are unchanged from Figure 15.4

Testing experiment E.3

We partially repeat experiment E.3 from increment 1 to show that the effects of altering the values of p and r in the input cistrome can be reproduced by increment 2. This experiment uses a synthetic, generated cistrome with $p = 1000$ (i.e. 1000 posts) and $r = 146$ (i.e. 146 red posts). The synthetic cistrome created is in effect a scaled down Nanog cistrome – for Nanog p=4310 and r=631.

The experiment starts with $s_0 = r$, that is, with all red posts active. The results are shown in Figure 18.2, and are unchanged from Figure 15.4 (right hand panel).

[Further details of testing omitted here.]

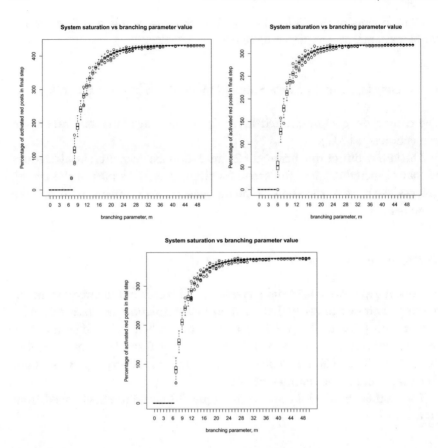

Fig. 18.1 Replication of E.1. Determination of the critical value of the branching parameter, m, for arenas constructed from the cistromes for the three core pluripotency transcription factors. The upper left hand panel shows the result for the Nanog arena, the
upper right for the Oct4 arena and the lower panel that for the Sox2 arena

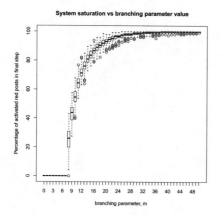

Fig. 18.2 Replication of E.3. $p = 1000$ and $r = 146$; $s_0 = r$ and m is varied from 0 to 50

Chapter 19
CellBranch: increment 2: Exploration Phase

Exploration Phase (97)
Use the Simulation Platform resulting from the Development Phase to explore the scientific questions established during the Discovery Phase.

- initially, revisit the Research Context (119)
- develop an experimental Results Model (174)
- finally, revisit the Simulation Purpose (121)

19.1 Exploration Phase > revisit Research Context

The Research Context is unchanged.

19.2 Exploration Phase > Results Model

Results Model (174)
Develop a results model suitable for interpreting Simulation Experiment data in Domain Model terms.

- perform Sensitivity Analysis (175)
- perform relevant Simulation Experiment (177)s
- build a Simulation Behaviours (179) model

© Springer Nature Switzerland AG 2018
S. Stepney, F.A.C. Polack, *Engineering Simulations as Scientific Instruments: A Pattern Language*, https://doi.org/10.1007/978-3-030-01938-9_19

19.2.1 Exploration Phase > Results Model > Sensitivity Analysis

No further Sensitivity Analysis (175) was performed

19.2.2 Exploration Phase > Results Model > Simulation Experiment

> **Simulation Experiment (177)** : Design, run, and analyse simulation experiments.

19.2.3 Exploration Phase > Results Model > Simulation Experiment > design

The design is unchanged from increment 1.

19.2.4 Exploration Phase > Results Model > Simulation Experiment > perform

Protocol

Each simulation run comprises the p and r values of a particular arena (chosen to match Nanog, Sox2, Oct4 data), with an m value and a starting activity as stated in each individual experiment.

For each simulation run, we record the proportion of active red posts at each timestep, producing a timeseries of the extent of cistrome activation.

For each parameter set (p, r, m, s_0), we run the simulation $N = 200$ times.

We identify several experiments to perform on the multiple-arena simulation:

Experiment E2.0

Preliminary investigation of system behaviour with the modified simulation. In each case the arenas have their critical value of the branching parameter as determined in increment 1.

- Simulate pairs of coupled arenas (Nanog/Oct4, Nanog/Sox2, Oct4/Sox2).
- Simulate a three coupled arena (Nanog/Sox2/Oct4).

E2.0	Nanog	Sox2	Oct4
r	631	542	466
m_c	8	7	6
m	m_c	m_c	m_c
s_0	r	r	r

Experiment E2.1

Test if an individually sustainable Oct4 arena can drive an initially totally dissipated Nanog arena.

E2.1	Nanog	Oct4
r	631	466
m_c	8	6
m	m_c	m_c
s_0	0	r

Experiment E2.1rev

Test if an individually sustainable Nanog arena can drive an initially totally dissipated Oct4 arena.

E2.1rev	Nanog	Oct4
r	631	466
m_c	8	6
m	m_c	m_c
s_0	r	0

Experiment E2.2

Test system behaviour when the Nanog arena has $m = 2 < m_c$, and the Oct4 arena has $m = 12 \gg m_c$. Both arenas have all red posts initially active.

E2.2	Nanog	Oct4
r	631	466
m_c	8	6
m	2	12
s_0	r	r

Experiment E2.3

Test different combinations of Nanog branching parameter, m, for the Nanog and Oct4 arenas, to see when critical Oct4 can ignite and maintain subcritical Nanog.

E2.3	Nanog	Oct4
r	631	466
m_c	8	6
m	5,6,7,8	m_c
s_0	0	r

Experiment E2.3rev

As for experiment E2.3, but with reversed initial conditions, to see when critical Nanog can ignite and maintain subcritical Oct4.

E2.3rev	Nanog	Oct4
r	631	466
m_c	8	6
m	m_c	3,4,5,6
s_0	r	0

Experiment E2.4

Test system behaviour when the Nanog arena has no branching, and the Oct4 arena has high branching, to see if Oct4 can sustain a non-branching Nanog. Both arenas have all red posts initially active.

E2.4	Nanog	Oct4
r	631	466
m_c	8	6
m	0	12
s_0	r	r

Experiment E2.5

Determine the minimum value of m for which sustainable Nanog arena can be ignited via combined activity in the Sox2 and Oct4 arenas.

E2.6	Nanog	Sox2	Oct4
r	631	542	466
m_c	8	7	6
m	4,5,6,7	m_c	m_c
s_0	0	r	r

Experiment E2.6

Determine the minimum value of m for which sustainable Oct4 arena can be ignited via combined activity in the Nanog and Sox2 arenas.

E2.7	Nanog	Sox2	Oct4
r	631	542	466
m_c	8	7	6
m	m_c	m_c	$0-m_c$
s_0	r	r	0

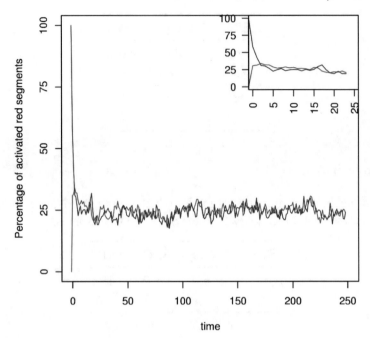

Fig. 19.1 E2.1: Oct4 igniting Nanog. A critically branching Oct4 cistrome (blue trace) with initial activation $s_0 = r$ and $m = m_c$ at time $t = 0$ drives the branching in an initially fully dissipated Nanog cistrome (red trace) with initial activation $s_0 = 0$ and $m = m_c$

19.2.5 Exploration Phase > Results Model > Simulation Experiment > analyse results

Experiment E2.0 results

All the possible arena pairings, and the three coupled areas, show similar behaviour with average activation around 25% for each arena throughout the duration of the simulation.

  [Plot omitted]

Experiment E2.1 results

The Oct4 cistrome critical branching process can drive an initially dissipated Nanog cistrome branching process, through activation of red posts that the two arenas share, Figure 19.1. The plot shows the first 250 of 1000 timesteps performed, and shows the first run from a set of 200 runs performed; the others runs are similar. The inset shows the first 25 timesteps.

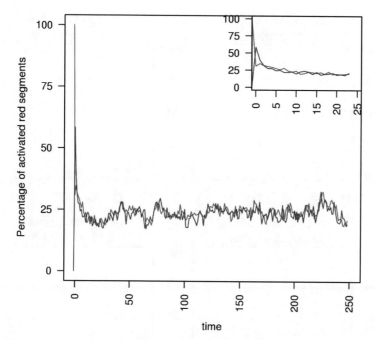

Fig. 19.2 E2.1rev: Nanog igniting Oct4. A critically branching Nanog cistrome (red trace) with initial activation $s_0 = r$ and $m = m_c$ at time $t = 0$ drives the branching in an initially fully dissipated Oct4 cistrome (blue trace) with initial activation $s_0 = 0$ and $m = m_c$

Experiment E2.1rev results

The Nanog cistrome critical branching process can drive an initially dissipated Oct4 cistrome branching process, through activation of red posts that the two arenas share, Figure 19.2. The plots shows the first 250 of 1000 timesteps performed, and shows the first run from a set of 200 runs performed; the others runs are similar. The inset shows the first 25 timesteps.

Experiment E2.2 results

A subcritically branching Nanog arena interacting with a supercritically branching Oct4 arena: if the branching parameter is high enough in the Oct4 arena it can still drive activity in the Nanog arena even with very low values of the branching parameter in this arena.

[Plot omitted]

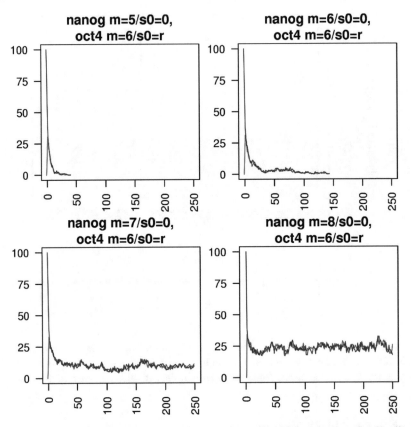

Fig. 19.3 E2.3: Oct4 igniting Nanog, varying m. The effect of coupling a subcritically or critically branching Nanog cistrome (red trace) to a critically branching Oct4 cistrome (blue trace). The m_c value in the Nanog cistrome is lowered from 8 to 7

Experiment E2.3 results

The branching process in coupled Oct4 and Nanog arenas is sustainable even if the m_c value in the Nanog cistrome is lowered from 8 to 7, Figure 19.3. The plots shows the first 250 of 1000 timesteps performed, and shows the first run from a set of 200 runs performed; the others runs are similar.

Experiment E2.3rev results

The branching process in coupled Oct4 and Nanog arenas is sustainable even if the m_c value in the Oct4 cistrome is lowered from 6 to 5, Figure 19.4.

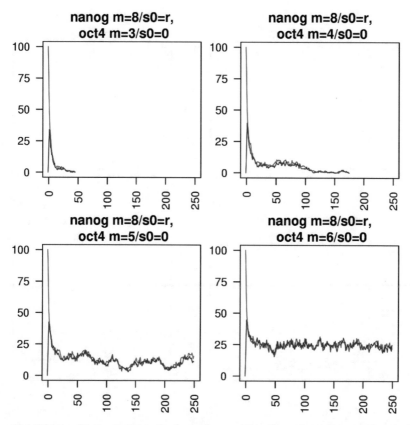

Fig. 19.4 E2.3rev: Nanog igniting Oct4, varying m. The effect of coupling a subcritically or critically branching Oct4 cistrome (blue trace) to a critically branching Nanog cistrome (red trace). The m_c value in the Oct4 cistrome is lowered from 6 to 5

The plots shows the first 250 of 1000 timesteps performed, and shows the first run from a set of 200 runs performed; the others runs are similar.

Experiment E2.4 results

The system behaviour when an initially fully activated, but non-branching Nanog arena ($s_0 = r$ and $m = 0$) interacts with a fully activated and super-critically branching Oct4 arena ($s_0 = r$ and $m = 12$, twice the value of m_c determined for the Oct4 arena) is comparable to that in Experiment E2.2. The Oct4 arena is still capable of driving short-lived activity in the Nanog cistrome, but this is all derived from the activation of shared red posts.

[Plot omitted]

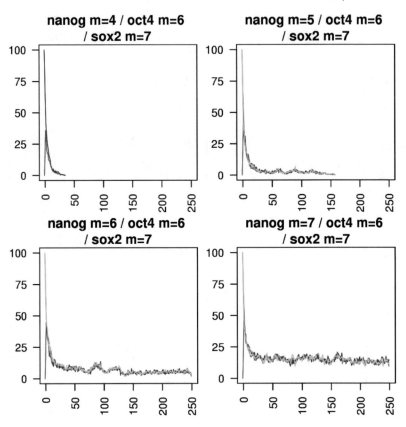

Fig. 19.5 E2.0: Coupling the Nanog arena to both the Oct4 and Sox2 arena. The effect of coupling a subcritically or critically branching Nanog arena (red trace) to a critically branching Oct4 arena (blue trace) and a critically branching Sox2 arena (green trace). The m_c value in the Nanog arena is lowered from 8 to 6

Experiment E2.5 results

Varying the value of the branching parameter, m, for the Nanog arena while it is coupled to critically branching Oct4 and Sox2 arenas reveals that the coupled Nanog arena can now sustain branching when $m = 6$, which represents an effective reduction of m_c from 8 to 6, Figure 19.5. The plots shows the first 250 of 1000 timesteps performed, and shows the first run from a set of 200 runs performed; the others runs are similar.

arena Y	arena X m_c	Nanog 8	Sox2 7	Oct4 6
Nanog	8	–	4.9	4.0 / **5**
Sox2	7	6.0 / **7**	–	4.6
Oct4	6	6.3 / **7**	5.6	–
Sox2+Oct4	7+6	**6**		

Fig. 19.6 Simulation Behaviours: Italic text shows the infinite limit Domain Behaviours predictions (reproduced from Figure 17.5 for comparison); bold text shows the (slightly higher) measured values for m_c for various couplings. The second cistrome Y is run at its original value of m_c. Sox2 and Oct4 can each decrease Nanog's m_c by 1, and working together by 2. Nanog can reduce Oct4's m_c by 1

Experiment E2.6 results

Varying the value of the branching parameter, m, for the Nanog arena while it is coupled to a critically branching Oct4 arena reveals that the coupled Nanog arena can now sustain branching when $m = 7$, which represents an effective reduction of m_c from 8 to 7.

[Plot omitted]

19.2.6 Exploration Phase > Results Model > Simulation Behaviours

Simulation Behaviours (179) : Develop a model of the emergent properties of a Simulation Experiment, for comparison with the related emergent Domain Behaviours of the Domain Model.

19.2.7 Exploration Phase > Results Model > Simulation Behaviours > minimal model

We observe that coupling cistromes permits an effective lowering of the critical value of the branching parameter, m_c. Coupling the Nanog arena to either the Oct4 or the Sox2 arena lowers its effective m_c by 1 if the other arena is branching at its critical rate m_c. Coupling the Nanog arena to both the Oct4 and Sox2 arenas, with both critically branching, lowers its effective m_c by 2, from 8 to 6. See Figure 19.6.

19.2.8 Exploration Phase > Results Model > Simulation Behaviours > Visualisation Model

- The visualisation employed in increment 1 mimics the single cistrome data in Figure 13.2. It is removed from increment 2, as it did not prove useful in increment 1.
- Some output data is presented as plots of activation at $T = 1000$ against m, for multiple arenas [these plots not shown]
- Some output data is presented as plots of activation against time, for multiple arenas

19.3 Exploration Phase > revisit Simulation Purpose

The purpose of the CellBranch simulation project to provide evidence of the usefulness of BPT as a model of decision making in stem cell differentiation.

The increment 2 Simulation Experiments demonstrate the effect of coupled arenas, and in particular that one arena can ignite and maintain another. This provides evidence in favour of BPT as a suitable model.

It is now time to decide how to progress this avenue of research.

Chapter 20
CellBranch: Lessons and code

20.1 CoSMoS lessons

This case study documents and illustrates the use of CoSMoS patterns to perform a CoSMoS simulation project, from initial discovery, through development, to exploration, over two increments. There were several lessons learned about the use of CoSMoS, which are summarised here.

It is not always clear whether information should be included in the Domain, or Domain Model, sections, particularly relating to assumptions. Similarly, some of the preliminary experiments to determine m_c might be considered to be Calibration, or Sensitivity Analysis. What is important, however, is to document the information, rather than to agonise over precisely which section to document it in.

Not all pattern material need be presented in the pattern order. For example, here the Cartoon for the Domain Model was best presented within the Basic Domain Model section (Figure 13.4), rather than as a prior illustration. Not all core patterns are applicable. For example, here the TF BP model is so abstracted from the Domain (123), that aspects such as the Domain Experiment Model are not relevant. It is more important to follow the spirit of the CoSMoS approach rather than the letter of every pattern.

Not every aspect of the CoSMoS approach needs to be performed with complete rigour. The CellBranch simulation is not safety critical, so some aspects have been omitted (such as justification of every assumption, and argumentation of fitness-for-purpose). The extra effort needed to complete all aspects should be expended only if it gives benefit.

Although the presentation is sequential and hierarchical, the historical process was not. We spent many short iterations, and considerable backtracking (for example, see Figure 13.3), before finally fixing on the 'sparking posts' model in increment 1. The CoSMoS patterns define what information

© Springer Nature Switzerland AG 2018
S. Stepney, F.A.C. Polack, *Engineering Simulations as Scientific Instruments: A Pattern Language*, https://doi.org/10.1007/978-3-030-01938-9_20

should be recorded by the end of the project, but not the order it needs to be produced. Some uses of CoSMoS can apply the patterns in significantly different orders, for example [12].

We might not have arrived at the conceptual sparking posts model without taking an incremental approach. The need to have just a single-cistrome model for this first increment revealed a fundamental misunderstanding that the modellers were having about the background TF BP model.

We were taking an agile approach, producing minimal simulation models and code, and collaboration meetings would often generate interesting but out of current scope ideas. We invented the concept of the !! **Future** tag (which we dubbed the "to don't" list): a way to record the ideas for future reference, in a manner that made it clear they were not to be included in the current increment. Some of these ideas also prompted the recognition of assumptions in the current increment.

The Domain Scientist (Halley) was new to the CoSMoS approach at the start of the project, but had previous experience working with modellers using different approaches on other projects. Halley reports that CoSMoS is a flexible tool to produce objective scientific simulations, and allows progress without being funnelled into preconceptions imposed by a specific toolset or implementation approach.

20.2 The Simulation Platform code

The code for the simulation, batch scripts for running the simulation on an SGE enabled compute cluster, Python scripts for generating real or synthetic cistromes, and example R scripts for processing simulation results into graphical form, are all available on GitHub at: github.com/CellBranch/CellBranch

Part V
Appendices

Endnotes

Preface

1 [p.vi] Recommended software engineering approaches include agile development [27], test-driven development [28], and use of a Domain Specific Language (228) [78].

2 [p.vi] Examples of possible modelling approaches and notations are Ordinary Differential Equations (ODEs), Agent Based Modelling (208), Petri nets, and the Unified Modeling Language (UML) [77].

3 [p.vii] Publications produced by the CoSMoS project team, and by associated projects, include:

- simulation, modelling and process descriptions [10, 15, 178, 179, 207]
- validation and argumentation [11, 88, 89, 176]
- environment orientation [117, 118, 120]
- metamodels [14, 119]
- other aspects of the CoSMoS approach [12, 13, 204]
- various case studies and uses of CoSMoS [1, 59, 63, 71, 85, 86, 98, 180, 187]
- CoSMoS workshop proceedings [205, 206, 208–213]

Many of these publications are available from the CoSMoS project website www.cosmos-research.org

© Springer Nature Switzerland AG 2018
S. Stepney, F.A.C. Polack, *Engineering Simulations as Scientific Instruments: A Pattern Language*, https://doi.org/10.1007/978-3-030-01938-9

Chapter 1. Introduction

1 [p.7] There are many incremental variants of the Sargent paper in a series of workshops. The 2005 version [201] is the relevant reference here.

2 [p.14] The advantages of practice over theory are recognised by Aristotle [16, book I, Chapter 2]:

> those who dwell in intimate association with nature and its phenomena grow more and more able to formulate, as the foundations of their theories, principles such as to admit of a wide and coherent development: while those whom devotion to abstract discussions has rendered unobservant of the facts are too ready to dogmatize on the basis of a few observations.

3 [p.17] By "assumption" we mean any kind of abstraction, simplification, axiom, idealisation or approximation.

4 [p.18] The difference between a descriptive (scientific) model and a prescriptive (engineering) model lies in the consequence of a mismatch between the model and what is modelled. In the descriptive case, the model needs to be changed to bring it closer to the real system. In the prescriptive case, the system needs to be changed (re-engineered) to bring it closer to the model (specification).

If the Domain is an engineered domain, rather than a natural domain, the Domain Model is then a prescriptive (engineering) model. However, it is still a distinct model from the prescriptive Platform Model.

5 [p.23] Macmillan online dictionary, www.macmillandictionary.com/dictionary/british/fit-for-purpose

Chapter 2. What's in it for me?

1 [p.37] When the robot needs to physically interact with (or avoid) another robot, the simulator does not need to predict trajectories with high accuracy since, as real world trajectories converge, the simulation accuracy on repeated simulation cycles increases simply because the robots are closer. Thus limited fidelity is good enough.

Chapter 3. The process in miniature

1 [p.43] The running example of cell division and differentiation in the prostate is based on work done on the AlKan project at York. A fuller description of the models can be found in [63], and of the fitness for purpose argument in [180]. The example presented herein differs in some details from that published work, for expository purposes.

Additionally, the running example captures only the first of a series of mini-projects exploring the simulation engineering and the biology of prostate cell proliferation; subsequent mini-projects revisited many of the decisions recorded here. In particular, in a later iteration of AlKan it was agreed that a much smaller (in terms of number of cells), 3D spatial simulator was needed, to match results to tissue slices that had been stained and imaged for "visual validation".

The AlKan work was supported by Program Grant support (to Prof. N. J. Maitland) from Yorkshire Cancer Research, by TRANSIT (EPSRC grant EP/F032749/1) through the York Centre for Complex Systems Analysis, and by CoSMoS (EPSRC grant EP/E053505/1). We thank the members of the Cancer Research Unit in York for their invaluable input to the project.

2 [p.63] A Petri net is "a formal, graphical, executable technique for the specification and analysis of concurrent, discrete-event dynamic systems" [126, 127]. A Petri net is a bipartite directed graph, with *place* nodes and *transition* nodes. Place nodes have a *marking*: a set of *tokens* occupying the place. A transition can *fire* when all its input places hold a token: one token per input arc is *consumed* from its input place, and one token per output arc is *produced* in the respective output. Petri nets naturally model concurrency, as multiple tokens can be traversing a net simultaneously.

Chapter 5. Discovery Phase Patterns

1 [p.124] We use the term Cartoon (124), or sketch, for a picture that has no defined syntax or semantics, no explicit Metamodel (234). The notation and meaning is partly *ad hoc*, domain-specific convention (which may be unknown to some of the development team), and partly tacit knowledge of the cartoonist and of the reader (this tacit knowledge may be different between

the parties). We reserve the term *diagram*, on the other hand, for a picture that at least has a well-defined syntax (which may be produced and checked by some tool), and may have a well-defined semantics (so it may be transformed into other forms, such as code fragments). Because of this extra precision and detail, diagrams can be harder for non-specialists to understand.

2 [p.129] The word 'domain' has a subtly different meaning in software engineering, where there is not such as strong separation between the scientific domain and the simulation of it. See, for example, Domain-Driven Design in [67], domain models in [76]. Crucially, the CoSMoS Domain Model includes the model of the hypothesis (for example, emergent properties), which should *not* appear in the implementation. Additionally, the Domain Model should be platform neutral.

3 [p.133] This calibration and validation data is analogous to the training and test data sets used in machine learning [92]. It is needed to ensure that the simulation is not so tuned that it "overfits" the calibration (training) data, but is instead generic enough to also fit the (unseen during calibration) validation data.

4 [p.136] Aevol is an *in silico* experimental artificial *evol*ution platform [24] in which populations of digital bacteria are subject to Darwinian-style evolution. Its Domain (123) falls within the areas of evolutionary theory and digital genetics focussing on the evolutionary dynamics of the size and organisation of bacterial genomes. The Aevol simulator encapsulates an *in silico* laboratory to test evolutionary scenarios, enabling simulation experiments in which populations of artificial organisms evolve within a controlled environment. These experiments mimic those used in real bacterial evolutionary studies, the most famous of which is the Lenski long-term evolutionary experiment [234]. Aevol provides many of the same experimental setup tools as this experimental system. Aevol was not developed using the CoSMoS approach, but a Domain Model, including a Domain Experiment Model, has been reversed engineered from the Aevol simulator [12].

5 [p.136] A specific example of this problem occurred with experiments performed in the mouse autoimmunity model Experimental Autoimmune Encephalomyelitis (EAE), a proxy for multiple sclerosis. These report results on a six point scale: 0, no symptoms; 1, flaccid tail; 2, hind limb weakness; 3, hind limb paralysis; 4, whole body paralysis; 5, death [145]. Read et al. [189] employed simulation as a tool to investigate the role of the immune system in the development of autoimmunity and subsequent recovery in these mice. However, no such observations could be made of a simulation, com-

plicating comparison between simulation and domain experiments, and calibration. Making the assumption that EAE disease severity scores were well-correlated with harm to neurons, abstracted in their simulation as neuronal death, Read et al. devised a metric to grade simulation autoimmunity in terms of this six point scale based on rates of neuronal death. This metric required calibration, and herein two domain experiments were employed, each independently reproduced in simulation: physiological recovery from disease induction in these mice, and an intervention that prolonged and worsened disease symptoms. Rates of simulated neuronal death, measured per hour, fluctuated with far higher frequency than changes in EAE severity score, which occurred on the order of days. The neuronal death time-series required smoothing. Thereafter a threshold value of neuronal death rate could be defined for each disease severity score. Smoothing parameters and thresholds were calibrated whilst taking consideration of multiple aspects of *in vivo* EAE progression. *In silico* disease scores needed to reproduce both the distribution of maximum disease severity scores attained by mice in each experiment, and the frequencies at which disease scores changed (attained through Fourier transformations). Completed, this "EAE-O-Meter" simulation metric facilitated direct comparison between simulation and domain results, and provided a language through which simulation behaviours could be interpreted in domain terms. Their methodology is fully reported in [185].

Chapter 6. Development Phase Patterns

1 [p.160] For examples of some such libraries or frameworks, see Chapter 9's endnote 2.

2 [p.162] Eclipse is a commonly used and powerful IDE and suite of development tools, available from www.eclipse.org

3 [p.166] Read et al.'s EAE-O-Meter (introduced in endnote 5 in the Domain Experiment Model (135) pattern in Chapter 5) was constructed only after the baseline simulation was calibrated. Simulation calibration was performed on the basis of cell population (relative) levels in several organs [185]. Compiling this data *in vivo* is difficult. The immune response is inherently dynamic, and disease progression, observed at the system level (the six point scale above) and also at the macro level (cell population numbers), vary

considerably between individuals. Longitudinal observations in single individuals would be ideal. However, collecting this data requires the sacri-
fice of an animal such that a single animal can provide only a single snap-
shot of the overall time-series process. (Ethical and) practical considerations
limit the number of animals employed in these studies, and given the inter-
individual variation, this presents a challenge to collecting high quality data
for simulation calibration. Read et al. performed their simulation calibration
against a Domain Scientist's intuition of how the system operates, itself build
upon years of absorbing piecemeal data such as this from the literature and
their own experiments.

4 [p.175] Sensitivity analysis is closely related to uncertainty analysis. The
former establishes the influence a system's inputs has on its responses,
whereas the latter explicitly focuses on what range of response values result from a range of input values. In this book, the term Sensitivity Analysis is
used to describe both techniques.

5 [p.177] We give here a specific example of Sensitivity Analysis conducted in a
CoSMoS Simulation Project.

Sensitivity analyses were conducted to understand an agent-based model
of the development of organs that trigger an adaptive immune response to
infection [4, 5]. Using a combination of local and global sensitivity analyses
techniques, specifically latin-hypercube sampling and applying the extended fourier amplitude test (eFAST) for the latter, it was suggested that the
key pathway in organ development identified in the literature was not influencing cell behaviour at simulated hour 12, yet was highly influential by
the end of organ development at hour 72. Although an unexpected result
that was contrary to that predicted when referencing the literature and in
discussions with the domain scientist, the application of the CoSMoS process provided confidence that the model was fit for the purpose for which it
was designed, and thus this unexpected result may not have been produced
due to an error in implementation. As such the Domain Scientist conducted
an experiment in the laboratory to test this hypothesis, specifically that adhesion factors were more influential than the expected chemokine factors at
hour 12, which produced cell behaviours that were in agreement with those
produced in simulation [174]. The interesting fact that the chemokine pathway was identified as influential in sensitivity analyses at the end of the
simulation, as expected in the literature, gave rise to the important question
of when the change in influential biological pathways occurs during organ
development. As such temporal sensitivity analyses were conducted at 12
hour intervals to examine the influence of adhesion and chemokine path-

ways over time, with the analyses suggesting that a change in influential biological pathway occurs between hours 24 and 36 of organ development [6]. Through the construction of the simulator, using principled design approaches described in the CoSMoS patterns, and detailed sensitivity analyses, it could be suggested that organ development was in fact bi-phasic, and more complex than that suggested in the academic literature.

Chapter 9. Modelling and documentation patterns

1 [p.209] "Individual-based modelling" and "agent-based modelling" are synonymous. The former term is sometimes preferred if the components do not have "agency"; that is, if they merely react. It is also used in the ecological domain. However, there is no sharp distinction in meaning in the literature. We use the term "agent" throughout this book, with no implication of "agency" or "intelligence".

2 [p.210] NetLogo is a multi-agent programmable modelling environment for prototyping multi-agent simulations. It provides a programming language and user-interface widgets to support rapid prototyping of relatively simple agent-based simulations. It has a large user-base and a large library of example simulations. See ccl.northwestern.edu/netlogo/.

MASON, from George Mason University, is a discrete-event multiagent simulation library, which can be used as the foundation for Java simulations. See cs.gmu.edu/~eclab/projects/mason/.

FLAME (Flexible Large-scale Agent Modelling Environment), from the University of Sheffield, is an agent-based modelling system. A model is defined as an extended finite state machine. From this FLAME generates a complete agent-based application, which can be targeted to computing systems ranging from laptops to supercomputers. See www.flame.ac.uk/. FLAME GPU (www.flamegpu.com/) is an extension to the FLAME framework optimised for GPUs.

Chapter 10. Real world simulation

1 [p.229] DSLs were originally dubbed "little-languages" [30], although they are not always that little.

2 [p.229] For more information on SBML, see [122] and the portal at sbml.org

References

1. Ali Afshar Dodson, Susan Stepney, Emma Uprichard and Leo Caves (2014). "Using the CoSMoS approach to study Schelling's Bounded Neighbourhood Model". *Proceedings of the 2014 Workshop on Complex Systems Modelling and Simulation, New York, NY, USA, July 2014*. Ed. by Susan Stepney and Paul S. Andrews. Luniver Press, pp. 1–12 (see pp. 88, 233, 327).

2. Bruce Alberts, Alexander Johnson, Julian Lewis, Martin Raff, Keith Roberts and Peter Walter (2008). *Molecular Biology of the Cell*. 5th edition. Garland Science (see p. 126).

3. Kieran Alden, Paul S. Andrews, Fiona A. C. Polack, Henrique Veiga-Fernandes, Mark C. Coles and Jon Timmis (2015). "Using argument notation to engineer biological simulations with increased confidence". *Journal of the Royal Society, Interface* 12(104):20141059 (see pp. 191, 192).

4. Kieran Alden, Mark Read, Jon Timmis, Paul S. Andrews, Henrique Veiga-Fernandes and Mark C. Coles (2013). "*Spartan*: A Comprehensive Tool for Understanding Uncertainty in Simulations of Biological Systems". *PLoS Comput Biol* 9(2):e1002916 (see pp. 177, 332).

5. Kieran Alden, Jon Timmis, Paul S. Andrews, Henrique Veiga-Fernandes and Mark C. Coles (2012). "Pairing experimentation and computational modeling to understand the role of tissue inducer cells in the development of lymphoid organs". *Frontiers in Immunology* 3:172 (see p. 332).

6. Kieran Alden, Jon Timmis, Paul S. Andrews, Henrique Veiga-Fernandes and Mark C. Coles (2017). "Extending and Applying Spartan to Perform Temporal Sensitivity Analyses for Predicting Changes in Influential Biological Pathways in Computational Models". *IEEE/ACM Transactions on Computational Biology and Bioinformatics* 14(2):431–442 (see pp. 177, 333).

7. Christopher Alexander, Sara Ishikawa, Murray Silverstein, Max Jacobson, Ingrid Fiksdahl-King and Shlomo Angel (1977). *A Pattern Language: towns, buildings, construction*. Oxford University Press (see p. 13).

8. Akram Alyass, Michelle Turcotte and David Meyre (2015). "From big data analysis to personalized medicine for all: challenges and opportunities". *BMC Medical Genomics* 8(1):33 (see p. 38).

9.   P. W. Anderson (1991). "Is Complexity Physics? Is It Science? What is It?" *Physics Today* 44(7):9 (see p. 262).

10. Paul S. Andrews, Fiona A. C. Polack, Adam T. Sampson, Susan Stepney and Jon Timmis (2010). *The CoSMoS Process, Version 0.1: A Process for the Modelling and Simulation of Complex Systems*. Tech. rep. YCS-2010-453. Department of Computer Science, University of York (see p. 327).

11. Paul S. Andrews, Fiona Polack, Adam T. Sampson, Jon Timmis, Lisa Scott and Mark Coles (2008). "Simulating biology: towards understanding what the simulation shows". *Proceedings of the 2008 Workshop on Complex Systems Modelling and Simulation, York, UK*. Ed. by Susan Stepney, Fiona Polack and Peter Welch. Luniver Press, pp. 93–123 (see pp. 111, 327).

12. Paul S. Andrews and Susan Stepney (2014). "Using CoSMoS to Reverse Engineer a Domain Model for Aevol". *Proceedings of the 2014 Workshop on Complex Systems Modelling and Simulation, New York, NY, USA, July 2014*. Ed. by Susan Stepney and Paul S. Andrews. Luniver Press, pp. 61–79 (see pp. 233, 324, 327, 330).

13. Paul S. Andrews and Susan Stepney (2015). "The CoSMoS Domain Experiment Model". *Proceedings of the 2015 Workshop on Complex Systems Modelling and Simulation, York, UK, July 2015*. Ed. by Susan Stepney and Paul S. Andrews. Luniver Press (see p. 327).

14. Paul S. Andrews, Susan Stepney, Tim Hoverd, Fiona A. C. Polack, Adam T. Sampson and Jon Timmis (2011). "CoSMoS process, models, and metamodels". *Proceedings of the 2011 Workshop on Complex Systems Modelling and Simulation, Paris, France*. Ed. by Susan Stepney, Peter Welch, Paul S. Andrews and Carl G. Ritson. Luniver Press, pp. 1–13 (see pp. 237, 243, 327).

15. Paul S. Andrews, Susan Stepney and Jon Timmis (2012). "Simulation as a Scientific Instrument". *Proceedings of the 2012 Workshop on Complex Systems Modelling and Simulation, Orleans, France*. Ed. by Susan Stepney, Paul S. Andrews and Mark N. Read. Luniver Press, pp. 1–10 (see p. 327).

16. Aristotle (1922). *On Generation and Corruption*. translation by H. H. Joachim, available from classics.mit.edu/Aristotle/gener_corr.html (see p. 328).

17. A. Ay and D. N. Arnosti (2011). "Mathematical modeling of gene expression: a guide for the perplexed biologist". *Critical Reviews in Biochemistry and Molecular Biology* 46(2):137–151 (see p. 259).

18. P. Bak and M. Paczuski (1993). "Why Nature is complex". *Physics World* 6(12):39–43 (see p. 262).

19. P. Bak and M. Paczuski (1995). "Complexity, contingency, and criticality". *Proceedings of the National Academy of Science USA* 92:6689–6696 (see p. 262).

20. P. Ball (1999). "Transitions still to be made". *Nature* 402:C73–C76 (see p. 262).

21. P. Ball (2001). *The Self-Made Tapestry*. Oxford University Press (see pp. 261, 262).

22. Wolfgang Banzhaf, Bert Baumgaertner, Guillaume Beslon, René Doursat, James A. Foster, Barry McMullin, Vinicius Veloso de Melo, Thomas Miconi, Lee Spector, Susan Stepney and Roger White (2016). "Defining and Simulating Open-Ended Novelty: Requirements, Guidelines, and Challenges". *Theory in Biosciences* 135(3):131–161. DOI: 10.1007/s12064-016-0229-7 (see pp. 216, 235).

23. Michael Batty (2005). *Cities and Complexity: understanding cities with cellular automata, agent-based models, and fractals*. MIT Press (see p. 209).

24. Bérénice Batut, David P. Parsons, Stephan Fischer, Guillaume Beslon and Carole Knibbe (2013). "*In silico* experimental evolution: a tool to test evolutionary scenarios". *BMC Bioinformatics* 14(Suppl 15):S11 (see pp. 136, 330).

25. Bernhard Bauer and James Odell (2005). "UML 2.0 and Agents: How to Build Agent-based Systems with the new UML". *Journal of Engineering Applications of Artificial Intelligence* 18:141–157 (see p. 113).

26. Kent Beck (1997). *Smalltalk Best Practice Patterns*. Prentice Hall (see p. 13).

27. Kent Beck (2000). *Extreme Programming Explained*. Addison-Wesley (see pp. 161, 241, 327).

28. Kent Beck (2003). *Test-Driven Development*. Addison-Wesley (see p. 327).

29. Mohamed Ben Belgacem, Bastien Chopard, Joris Borgdorff, Mariusz Mamoński, Katarzyna Rycerz and Daniel Harezlak (2013). "Distributed Multiscale Computations Using the MAPPER Framework". *Procedia Computer Science* 18:1106–1115 (see p. 244).

30. Jon Bentley (1986). "Little languages". *Comms ACM* 29(8):711–721 (see p. 334).

31. Ann Blandford, Jo Gibbs, Nikki Newhouse, Olga Perski, Aneesha Singh and Elizabeth Murray (2018). "Seven lessons for interdisciplinary research on interactive digital health interventions". *Digital Health* 4:2055207618770325 (see p. 241).

32. Christian Blum, Alan F. T. Winfield and Verena V. Hafner (2018). "Simulation-Based Internal Models for Safer Robots". *Frontiers in Robotics and AI* 4:74 (see p. 37).

33. Eric Bonabeau (2002). "Agent-based modeling: Methods and techniques for simulating human systems". *PNAS* 99(Suppl 3):7280–7287 (see p. 209).

34. S. Bornholdt (2005). "Less is more in modeling large genetic networks". *Science* 310:449–451 (see p. 249).

35. Pierre Boutillier, Jérôme Feret, Jean Krivine and Lý Kim Quyên (2018). *Kappa tools reference manual*. v4.0. KappaLanguage.org. URL: github.com/Kappa-Dev/KaSim/releases (see p. 113).

36. Laurie A. Boyer, Tong Ihn Lee, Megan F. Cole, Sarah E. Johnstone, Stuart S. Levine, Jacob P. Zucker, Matthew G. Guenther, Roshan M. Kumar, Heather L. Murray, Richard G. Jenner, David K. Gifford, Douglas A. Melton, Rudolf Jaenisch and Richard A. Young (2005). "Core transcriptional regulatory circuitry in human embryonic stem cells". *Cell* 122:947–956 (see pp. 257, 261).

37. D. Bray (2003). "Molecular networks: the top-down view". *Science* 301:1864–1865 (see p. 249).

38. Florence Broders-Bondon, Thanh Huong Nguyen Ho-Bouldoires, Maria Elena Fernandez-Sanchez and Emmanuel Farge (2018). "Mechanotransduction in tumor progression: The dark side of the force". *Journal of Cell Biology* :DOI: 10.1083/jcb.201701039 (see p. 32).

39. William J. Brown, Raphael C. Malveau, Hays W. "Skip" McCormick III and Thomas J. Mowbray (1998). *AntiPatterns: refactoring software, architectures, and projects in crisis*. Wiley (see pp. 13, 15).

40. Alan Burns and Ian J. Hayes (2010). "A Timeband Framework for Modelling Real-Time Systems". *Real-Time Systems Journal* 45(1-2):106–142 (see p. 244).

41. "Novel Approaches to the Visualization and Quantification of Biological Simulations by Emulating Experimental Techniques" (2014). *ALife 14, New York, NY, USA, July 2014*. Ed. by James A. Butler, Kieran Alden, Henrique Veiga Fernandes, Jon Timmis and Mark Coles. MIT Press (see p. 182).

42. Bill Buxton (2007). *Sketching User Experiences: getting the design right and the right design*. Morgan Kaufmann (see p. 125).

43. Michael D Byrne (2013). "How many times should a stochastic model be run? An approach based on confidence intervals". *The 12th International Conference on Cognitive Modelling* (see p. 178).

44. Muffy Calder, Claire Craig, Dave Culley, Richard de Cani, Christl A. Donnelly, Rowan Douglas, Bruce Edmonds, Jonathon Gascoigne, Nigel Gilbert, Caroline Hargrove, Derwen Hinds, David C. Lane, Dervilla Mitchell, Giles Pavey, David Robertson, Bridget Rosewell, Spencer Sherwin, Mark Walport and Alan Wilson (2018). "Computational modelling for decision-making: where, why, what, who and how". *Royal Society Open Science* 5(6):(see p. v).

45. Cancer Research UK (2015). *Prostate cancer statistics – Prostate cancer incidence*. www.cancerresearchuk.org/health-professional/cancer-statistics/statistics-by-cancer-type/prostate-cancer. accessed 15/5/2018 (see p. 46).

46. L. C. Cantley, K. R. Auger, C. Carpenter, B. Duckworth, A. Graziani, R. Kapeller and S. Soltoff (1991). "Oncogenes and signal transduction". *Cell* 64(2):281–302 (see p. 46).

47. Pedro Casado, Edmund H. Wilkes, Farideh Miraki-Moud, Marym Mohammad Hadi, Ana Rio-Machin, Vinothini Rajeeve, Rebecca Pike, Sameena Iqbal, Santiago Marfa, Nicholas Lea, Steven Best, John Gribben, Jude Fitzgibbon and Pedro R. Cutillas (2018). "Proteomic and genomic integration identifies kinase and differentiation determinants of kinase inhibitor sensitivity in leukemia cells". *Leukemia* :DOI: 10.1038/s41375-018-0032-1 (see p. 38).

48. E. Chaisson (2004). "Complexity: An energetics agenda". *Complexity* 9(3):14–21 (see p. 261).

49. Claudine Chaouiya (2007). "Petri net modelling of biological networks". *Briefings in Bioinformatics* 8(4):210–219 (see p. 63).

50. Peter Checkland (1981). *Systems Thinking, Systems Practice: a 30-year retrospective*. Wiley (see p. 126).

51. Peter Checkland and Jim Scholes (1990). *Soft Systems Methodology in Action: a 30-year retrospective*. Wiley (see p. 126).

52. Curtis R. Chong and Pasi A. Jänne (2013). "The quest to overcome resistance to EGFR-targeted therapies in cancer". *Nature Medicine* 19:1389–1400 (see p. 35).

53. Bastien Chopard, Joris Borgdorff and Alfons G. Hoekstra (2014). "A framework for multi-scale modelling". *Philosophical transactions. Series A, Mathematical, physical, and engineering sciences* 372:20130378 (see p. 244).

54. Irun R. Cohen (2007). "Modeling immune behavior for experimentalists". *Immunological Reviews* 216(1):232–236 (see p. 26).

55. Anne T. Collins and Norman J. Maitland (2006). "Prostate cancer stem cells". *European Journal of Cancer* 42(9):1213–1218 (see p. 55).

56. James O. Coplien and Douglas C. Schmidt, eds. (1995). *Pattern Languages of Program Design*. Addison-Wesley (see p. 13).

57. J. P. Crutchfield, J. D. Farmer, Norman H. Packard and Robert S. Shaw (1986). "Chaos". *Scientific American* 255(6):38–49 (see p. 262).

58. Marija Cvijovic, Joachim Almquist, Jonas Hagmar, Stefan Hohmann, Hans-Michael Kaltenbach, Edda Klipp, Marcus Krantz, Pedro Mendes, Sven Nelander, Jens Nielsen, Andrea Pagnani, Natasa Przulj, Andreas Raue, Jörg Stelling, Szymon Stoma, Frank Tobin, Judith A. H. Wodke, Riccardo Zecchina and Mats Jirstrand (2014). "Bridging the gaps in systems biology". *Molecular Genetics and Genomics* 289(5):727–734 (see p. 39).

59. Keith De'Bell (2015). "Towards a Network Model of Community Empowerment for Public Health Outcomes: Application of the CoSMoS Approach to Social System Modelling". *Proceedings of the 2015 Workshop on Complex Systems Modelling and Simulation, York, UK, July 2015*. Ed. by Susan Stepney and Paul S. Andrews. Luniver Press, pp. 9–29 (see p. 327).

60. Ezequiel A. Di Paolo, Jason Noble and Seth Bullock (2000). "Simulation Models as Opaque Thought Experiments". *Artificial Life VII*. MIT Press, pp. 497–506 (see pp. 4, 6).

61. Marco Dorigo and Mauro Birattari (2007). "Swarm intelligence". *Scholarpedia* 2(9):1462 (see p. 240).

62. M. E. Driscoll (2009). *Is Big Data at a tipping point?* www.analyticbridge.com/profiles/blogs/is-big-data-at-a-tipping-point. [Accessed: 2015-05-01] (see p. 249).

63. Alastair Droop, Philip Garnett, Fiona A. C. Polack and Susan Stepney (2011). "Multiple model simulation: modelling cell division and differentiation in the prostate". *Proceedings of the 2011 Workshop on Complex Systems Modelling and Simulation, Paris, France*. Ed. by Susan Stepney, Peter Welch, Paul S. Andrews and Carl G. Ritson. Luniver Press, pp. 79–111 (see pp. 64, 66, 73, 74, 87, 212, 215, 245, 327, 329).

64. I. Ekeland (2002). "In the balance". *Nature* 417:385 (see p. 262).

65. Joshua M. Epstein (2008). "Why Model?" *Journal of Artificial Societies and Social Simulation* 11(4):12. ISSN: 1460-7425. URL: jasss.soc.surrey.ac.uk/11/4/12.html (see pp. 6, 121).

66. Joshua M. Epstein and Robert Axtell (1996). *Growing Artificial Societies: social science from the bottom up*. MIT Press (see p. 209).

67. Eric Evans (2004). *Domain-Driven Design: tackling complexity in the heart of software*. Addison-Wesley (see p. 330).

68. J. D. Farmer and N. H. Packard (1986). "Evolution, Games, and Learning: Models for Adaptation in Machines and Nature. An introduction to the Proceedings of the CNLS Conference, Los Alamos, May 1985". *Physica D* 22:vii–xii (see p. 262).

69. J. Ferrell (2009). "Q&A: Systems biology". *Journal of Biology* 28:(see pp. 249, 260).

70. S. Flowers (1996). *Software Failure: Management Failure: Amazing Stories and Cautionary Tales*. Wiley (see p. 27).

71. Anton Jakob Flügge, Jon Timmis, Paul Andrews, John Moore and Paul Kaye (2009). "Modelling and Simulation of Granuloma Formation in Visceral Leishmaniasis". *CEC 2009*. IEEE Press, pp. 3052–3059 (see p. 327).

72. Andrew Ford (2010). *Modeling the Environment*. 2nd edition. Island Press (see p. 126).

73. John Forrester, Richard Greaves, Howard Noble and Richard Taylor (2014). "Modeling social-ecological problems in coastal ecosystems: A case study". *Complexity* 19(6):73–82 (see p. 242).

74. Martin Fowler (1997). *Analysis Patterns: reusable object models*. Addison-Wesley (see p. 13).

75. Martin Fowler (1999). *Refactoring: improving the design of existing code*. Addison-Wesley (see p. 234).

76. Martin Fowler (2003). *Patterns of Enterprise Application Architecture*. Addison-Wesley (see p. 330).

77. Martin Fowler (2004). *UML Distilled: brief guide to the standard object modeling language*. 3rd edition. Addison-Wesley (see p. 327).

78. Martin Fowler (2011). *Domain-Specific Languages*. Addison-Wesley (see pp. 229, 327).

79. Roman Frigg and Stephan Hartmann (2009). "Models in Science". *The Stanford Encyclopedia of Philosophy*. Ed. by Edward N. Zalta. Summer 2009 edition. plato.stanford.edu/archives/sum2009/entries/models-science (see p. 5).

80. A. Füzéry, J. Levin, M. M. Chan and D. W. Chan (2013). "Translation of proteomic biomarkers into FDA approved cancer diagnostics: issues and challenges". *Clinical Proteomics* 10:13–26 (see p. 32).

81. Richard P. Gabriel (1996). *Patterns of Software: tales from the software community*. Oxford University Press (see p. 13).

82. R. Gallagher and T. Appenzeller (1999). "Beyond Reductionism". *Science* 284(5411):79 (see p. 267).

83. Antony Galton (2015). "Outline of a Formal Theory of Processes and Events, and Why GIScience Needs One". *International Workshop on Spatial Information Theory*. Vol. 9368. LNCS. Springer, pp. 3–22 (see pp. 6, 121).

84. Erich Gamma, Richard Helm, Ralph E. Johnson and John Vlissides (1995). *Design Patterns: elements of reusable object-oriented software*. Addison-Wesley (see pp. 13, 14).

85. Philip Garnett, Susan Stepney, Francesca Day and Ottoline Leyser (2010). "Using the CoSMoS Process to Enhance an Executable Model of Auxin Transport Canalisation". *Proceedings of the 2010 Workshop on Complex Systems Modelling and Simulation, Odense, Denmark*. Ed. by Susan Stepney, Peter H. Welch, Paul S. Andrews and Adam T. Sampson. Luniver Press, pp. 9–32 (see p. 327).

86. Philip Garnett, Susan Stepney and Ottoline Leyser (2008). "Towards an Executable Model of Auxin Transport Canalisation". *Proceedings of the 2008 Workshop on Complex Systems Modelling and Simulation, York, UK*. Ed. by Susan Stepney, Fiona Polack and Peter Welch. Luniver Press, pp. 63–91 (see p. 327).

87. Brian Gerkey, Richard T. Vaughan and Andrew Howard (2003). "The Player/Stage Project: Tools for Multi-Robot and Distributed Sensor Systems". *Proceedings of the 11th International Conference on Advanced Robotics (ICAR 2003)*, pp. 317–323 (see pp. 9, 30, 37).

88. Teodor Ghetiu, Robert D. Alexander, Paul S. Andrews, Fiona A. C. Polack and James Bown (2009). "Equivalence Arguments for Complex Systems Simulations – A Case-Study". *Proceedings of the 2009 Workshop on Complex Systems Modelling and Simulation, York, UK*. Ed. by Susan Stepney, Peter H. Welch, Paul S. Andrews and Jon Timmis. Luniver Press, pp. 101–140 (see p. 327).

89. Teodor Ghetiu, Fiona A. C. Polack and James L. Bown (2010). "Argument-Driven Validation of Computer Simulations – A Necessity Rather Than an Option". *VALID 2010:* IEEE Press, pp. 1–4 (see p. 327).

90. Nigel Gilbert (2008). *Agent-Based Models*. Sage Publications (see p. 209).

91. H. Randy Gimblett, ed. (2002). *Integrating Geographic Information Systems and Agent-based Modeling Techniques for Simulating Social and Ecological Processes*. Oxford University Press (see p. 212).

92. Paolo Giudici (2003). *Applied Data Mining: Statistical Methods for Business and Industry*. Wiley (see p. 330).

93. P. P. Glory, N. G. David and J. D. Emerald (2010). "Petri net models and non linear genetic diseases". *Bio-Inspired Computing: Theories and Applications (BIC-TA), 2010*. IEEE, pp. 1466–1470 (see p. 63).

94. J. P. Gollub and J. S. Langer (1999). "Pattern formation in nonequilibrium physics". *Reviews of Modern Physics* 71(2):S396–S403 (see p. 261).

95. Government Office for Science (2018). *Computational Modelling: Technological Futures*. www.gov.uk/government/publications/computational-modelling-blackett-review (see p. v).

96. Ann Grand, Clare Wilkinson, Karen Bultitude and Alan F. T. Winfield (2012). "Open Science: A new 'trust technology'?" *Science Communication* 34(5):679–689 (see p. 40).

97. Gran-Turisimo (n.d.). *GT Academy Winer Makes it to the Podium of Le Mans*. www.gran-turismo.com/hk/news/d13235.html. accessed 30/01/2012 (see p. 9).

98. Richard B. Greaves, Sabine Dietmann, Austin Smith, Susan Stepney and Julianne D. Halley (2015). "Genome-wide mouse embryonic stem cell regulatory network self-organisation: a big data CoSMoS computational modelling approach". *Proceedings of the 2015 Workshop on Complex Systems Modelling and Simulation, York, UK, July 2015*. Ed. by Susan Stepney and Paul S. Andrews. Luniver Press, pp. 31–66 (see pp. 242, 248, 250, 327).

99. Richard B. Greaves, Sabine Dietmann, Austin Smith, Susan Stepney and Julianne D. Halley (2016). *CellBranch CoSMoS model : increments 1, 2 and 3*. Tech. rep. YCS-2016-501. Department of Computer Science, University of York (see pp. 248, 250).

100. Richard B. Greaves, Sabine Dietmann, Austin Smith, Susan Stepney and Julianne D. Halley (2017). "A conceptual and computational framework for modelling and understanding the nonequilibrium gene regulatory networks of mouse embryonic

stem cells". *PLOS Computational Biology* 13(9):e1005713. DOI: 10.1371/journal.pcbi. 1005713 (see pp. 248, 250).

101. Richard B. Greaves, Fiona A. C. Polack and John Forrester (2012). "CoSMoS in the Context of Social-Ecological Systems Research". *Proceedings of the 2012 Workshop on Complex Systems Modelling and Simulation, Orleans, France*. Ed. by Susan Stepney, Paul S. Andrews and Mark N. Read. Luniver Press, pp. 47–76 (see p. 242).

102. Volker Grimm (2002). "Visual Debugging: A Way Of Analyzing, Understanding and Communicating Bottom-Up Simulation Models in Ecology". *Natural Resource Modeling* 15(1):23–38. DOI: 10.1111/j.1939-7445.2002.tb00078.x (see p. 168).

103. Volker Grimm, Uta Berger, Finn Bastiansen, Sigrunn Eliassen, Vincent Ginot, Jarl Giske, John Goss-Custard, Tamara Grand, Simone K. Heinz, Geir Huse, Andreas Huth, Jane U. Jepsen, Christian Jørgensen, Wolf M. Mooij, Birgit Müller, Guy Pe'er, Cyril Piou, Steven F. Railsback, Andrew M. Robbins, Martha M. Robbins, Eva Rossmanith, Nadja Rüger, Espen Strand, Sami Souissi, Richard A. Stillman, Rune Vabø, Ute Visser and Donald L. DeAngelis (2006). "A standard protocol for describing individual-based and agent-based models". *Ecological Modelling* 198(1-2) :115–126. DOI: 10.1016/j.ecolmodel.2006.04.023 (see pp. 8, 222, 226).

104. Volker Grimm, Uta Berger, Donald L. DeAngelis, J. Gary Polhill, Jarl Giske and Steven F. Railsback (2010). "The ODD protocol: A review and first update". *Ecological Modelling* 221(23):2760–2768. DOI: 10.1016/j.ecolmodel.2010.08.019 (see pp. 8, 222, 224).

105. Volker Grimm and Steven F. Railsback (2005). *Individual-based Modeling and Ecology*. Princeton University Press (see pp. 209, 222, 224).

106. Volker Grimm, Eloy Revilla, Uta Berger, Florian Jeltsch, Wolf M Mooij, Steven F Railsback, Hans-Hermann Thulke, Jacob Weiner, Thorsten Wiegand and Donald L DeAngelis (2005). "Pattern-oriented modeling of agent-based complex systems: lessons from ecology". *Science* 310(5750):987–991 (see p. 139).

107. *GSN Community Standard, v1* (2011). www.goalstructuringnotation.info. Origin Consulting (York) Limited, on behalf of the Contributors (see pp. 191, 198, 200).

108. J. D. Halley, F. R. Burden and D. A. Winkler (2009). "Stem cell decision making and critical-like exploratory networks". *Stem Cell Research* 2(3):165–177 (see p. 250).

109. J. D. Halley, K. Smith-Miles et al. (2012). "Self-organizing circuitry and emergent computation in mouse embryonic stem cells". *Stem Cell Research* 8(2):324–333 (see pp. 250, 254, 261, 294, 295).

110. J. D. Halley and D. A. Winkler (2008). "Critical-like self-organization and natural selection: Two facets of a single evolutionary process?" *BioSystems* 92(2):148–158 (see p. 250).

111. Roger Harrabin (2010). "Climate science must be more open, say MPs" :news.bbc. co.uk/1/hi/sci/tech/8595483.stm (see p. 7).

112. Neil B. Harrison, Brian Foote and Hans Rohnert, eds. (2000). *Pattern Languages of Program Design 4*. Addison-Wesley (see p. 13).

113. Leland H. Hartwell, John J. Hopfield, Stanislas Leibler and Andrew W. Murray (1999). "From molecular to modular cell biology". *Nature* 402:C47–C52 (see p. 249).

114. Alfons Hoekstra, Bastien Chopard and Peter Coveney (2014). "Multiscale modelling and simulation: a position paper". *Philosophical transactions. Series A, Mathematical, physical, and engineering sciences* 372:20130377 (see p. 244).

115. Owen E. Holland (2003). *Machine Consciousness*. Imprint Academic (see p. 239).

116. Leroy Hood (2014). "Systems Biology and P4 Medicine: Past, Present, and Future". *Rambam Maimonides Medical Journal* 4(2):(see p. 36).

117. Tim Hoverd and Adam T. Sampson (2010). "A Transactional Architecture for Simulation". *ICECCS 2010: Fifteenth IEEE International Conference on Engineering of Complex Computer Systems*. IEEE Press, pp. 286–290 (see p. 327).

118. Tim Hoverd and Susan Stepney (2009). "Environment orientation: an architecture for simulating complex systems". *Proceedings of the 2009 Workshop on Complex Systems Modelling and Simulation, York, UK*. Ed. by Susan Stepney, Peter H. Welch, Paul S. Andrews and Jon Timmis. Luniver Press, pp. 67–82 (see p. 327).

119. Tim Hoverd and Susan Stepney (2011). "Energy as a driver of diversity in open-ended evolution". *ECAL 2011, Paris, France, August 2011*. MIT Press (see p. 327).

120. Tim Hoverd and Susan Stepney (2015). "Environment Orientation: a structured simulation approach for agent-based complex system". *Natural Computing* 14(1):83–97. DOI: 10.1007/s11047-014-9449-2 (see pp. 211, 327).

121. Doug Howe, Maria Costanzo, Petra Fey, Takashi Gojobori, Linda Hannick, Winston Hide, David P Hill, Renate Kania, Mary Schaeffer, Susan St Pierre, Simon Twigger, Owen White and Seung Yon Rhee (2008). "Big data: The future of biocuration". *Nature* 455(7209):47–50 (see p. 249).

122. M. Hucka, A. Finney, H. M. Sauro, H. Bolouri, J. C. Doyle, H. Kitano, A. P. Arkin, B. J. Bornstein, D. Bray, A. Cornish-Bowden, A. A. Cuellar, S. Dronov, E. D. Gilles, M. Ginkel, V. Gor, I. I. Goryanin, W. J. Hedley, T. C. Hodgman, J.-H. Hofmeyr, P. J. Hunter, N. S. Juty, J. L. Kasberger, A. Kremling, U. Kummer, N. Le Novère, L. M. Loew, D. Lucio, P. Mendes, E. Minch, E. D. Mjolsness, Y. Nakayama, M. R. Nelson, P. F. Nielsen, T. Sakurada, J. C. Schaff, B. E. Shapiro, T. S. Shimizu, H. D. Spence, J. Stelling, K. Takahashi, M. Tomita, J. Wagner, J. Wang and the rest of the SBML Forum (2003). "The systems biology markup language (SBML): a medium for representation and exchange of biochemical network models". *Bioinformatics* 19(4):524–531 (see p. 334).

123. Paul Humphreys (2004). *Extending Ourselves: Computational Science, Empiricism, and Scientific Method*. Oxford University Press (see pp. 5, 7).

124. Yoshinori Imamura, Toru Mukohara, Yohei Shimono, Yohei Funakoshi, Naoko Chayahara, Masanori Toyoda, Naomi Kiyota, Shintaro Takao, Seishi Kono, Tetsuya Nakatsura and Hironobu Minami (2015). "Comparison of 2D- and 3D-culture models as drug-testing platforms in breast cancer". *Oncology Reports* 33:1837–1843 (see p. 32).

125. Darrel C. Ince, Leslie Hatton and John Graham-Cumming (2012). "The case for open computer programs". *Nature* 482(7386):485–488 (see p. 7).

126. *High-level Petri Nets - Concepts, Definitions and Graphical Notation* (2000). International Standard ISO/IEC 15909. Final Committee Draft: www.petrinets.info/docs/pnstd-4.7.4.pdf (see p. 329).

127. *Software and Systems Engineering – High-level Petri Nets Part 2: Transfer Format* (2005). International Standard ISO/IEC 15909. WD 15909-2:2005(E): www.petrinets. info/docs/ISO-IEC15909-2.WD.V0.9.0.pdf (see p. 329).

128. N. Jacobi, P. Husbands and I. Harvey (1995). "Noise and the reality gap: The use of simulation in evolutionary robotics". *Proceedings of the third European conference on Advances in Artificial Life*. Springer, pp. 704–720 (see p. 30).

129. A. Jain, J. Balaram, J. Cameron, J. Guineau, C. Lim, M. Pomerantz and G. Sohl (2004). "Recent developments in the ROAMS planetary rover simulation environment". *Proceedings IEEE Aerospace Conference 2004*. Vol. 2. IEEE, pp. 861–876 (see p. 10).

130. L. P. Kadanoff (1987). "Chaos: A View of Complexity in the Physical Sciences". *From Order to Chaos II Essays: Critical Chaotic and Otherwise*. World Scientific (see p. 262).

131. T. Kalkan and A. Smith (2014). "Mapping the route from naive pluripotency to lineage specification". *Phil. Trans. R. Soc. B* 369:20130540 (see p. 262).

132. Sergey Karabasov, Dmitry Nerukh, Alfons Hoekstra, Bastien Chopard and Peter V. Coveney (2014). "Multiscale modelling: approaches and challenges". *Philosophical transactions. Series A, Mathematical, physical, and engineering sciences* 372:20130390 (see p. 244).

133. Aleksandra Karolak, Dmitry A. Markov, Lisa J. McCawley and Katarzyna A. Rejniak (2018). "Towards personalized computational oncology: from spatial models of tumour spheroids, to organoids, to tissues". *Journal of The Royal Society Interface* 15(138):DOI: 10.1098/rsif.2017.0703 (see pp. 33, 36).

134. S. Kauffman (1995). *At Home in the Universe*. Oxford University Press (see p. 262).

135. Daniel A Keim, Florian Mansmann, Jörn Schneidewind, Jim Thomas and Hartmut Ziegler (2008). "Visual Analytics: Scope and Challenges". *Visual Data Mining*. Ed. by Simeon J Simoff, Michael H Böhlen and Arturas Mazeika. LNCS. Springer, pp. 76–90 (see p. 181).

136. Tim P. Kelly (1999). "Arguing safety – a systematic approach to managing safety cases". PhD thesis. Department of Computer Science, University of York (see p. 191).

137. Joshua Kerievsky (2005). *Refactoring to Patterns*. Addison-Wesley (see p. 234).

138. Jonghwan Kim, Jianlin Chu, Xiaohua Shen, Jianlong Wang and Stuart H. Orkin (2008). "An Extended Transcriptional Network for Pluripotency of Embryonic Stem Cells". *Cell* 132(6):1049–1061 (see p. 263).

139. Lauren Kimlin, Jareer Kassis and Victoria Virador (2013). "3D In Vitro Tissue Models and Their Potential for Drug Screening". *Expert Opinion on Drug Discovery* 8(12):1455–1466 (see p. 32).

140. Hiroaki Kitano (2002a). "Computational systems biology". *Nature* 420(6912):206–210 (see pp. 4, 39).

141. Hiroaki Kitano (2002b). "Systems biology: a brief overview". *Science* 295(5560):1662–1664 (see p. 4).

142. Anneke Kleppe, Jos Warmer and Wim Bast (2003). *MDA Explained: the Model Driven Architecture: practice and promise.* Addison-Wesley (see p. 235).

143. Andrew Koenig (1995). "Patterns and Antipatterns". *Journal of Object-Oriented Programming* 8(1):46–48 (see pp. 13, 15).

144. Pamela K. Kreeger and Douglas A. Lauffenburger (2010). "Cancer systems biology: a network modeling perspective". *Carciogenesis* 31(1):2–8 (see p. 39).

145. V. Kumar and E. E. Sercarz (1993). "The involvement of T cell receptor peptide-specific regulatory CD4+ T cells in recovery from antigen-induced autoimmune disease." *Journal of Experimental Medicine* 178(3):909–916 (see p. 330).

146. A. D. Lander (2010). "The edges of understanding". *BMC Biology* 8(1):40 (see p. 249).

147. Timothy R. Levine, René Weber, Craig Hullett, Hee Sun Park and Lisa L. Massi Lindsey (2008). "A Critical Assessment of Null Hypothesis Significance Testing in Quantitative Communication Research". *Human communication research* 34(2):171–187 (see p. 180).

148. Yuin-Han Loh, Qiang Wu, Joon-Lin Chew, Vinsensius B Vega, Weiwei Zhang, Xi Chen, Guillaume Bourque, Joshy George, Bernard Leong, Jun Liu, Kee-Yew Wong, Ken W Sung, Charlie W H Lee, Xiao-Dong Zhao, Kuo-Ping Chiu, Leonard Lipovich, Vladimir A Kuznetsov, Paul Robson, Lawrence W Stanton, Chia-Lin Wei, Yijun Ruan, Bing Lim and Huck-Hui Ng (2006). "The Oct4 and Nanog transcription network regulates pluripotency in mouse embryonic stem cells". *Nature Genetics* 38(4):431–440 (see p. 264).

149. B. D. MacArthur, A. Ma'ayan and I. R. Lemischka (2008). "Toward Stem Cell Systems Biology: From Molecules to Networks and Landscapes". *Cold Spring Harbor Symposia on Quantitative Biology* 73:211–215 (see pp. 266, 267).

150. Roy MacLean, Susan Stepney, Simon Smith, Nick Tordoff, David Gradwell, Tim Hoverd and Simon Katz (1994). *Analysing Systems: determining requirements for object-oriented development.* Prentice Hall (see pp. 126, 231).

151. Miles MacLeod and Nancy J. Nersessian (2013). "Building Simulations from the Ground Up: Modeling and Theory in Systems Biology". *Philosophy of Science* 80(4):533–556 (see p. 3).

152. Stewart Mader (2008). *Wikipatterns.* Wiley (see p. 220).

153. Norman J. Maitland and Anne T. Collins (2005). "A tumour stem cell hypothesis for the origins of prostate cancer". *BJU International* 96(9):1219–1223 (see pp. 43, 56).

154. Hugo Gravato Marques and Owen E. Holland (2009). "Architectures for functional imagination". *Neurocomputing* 72(4–6):743–759 (see p. 240).

155. G. Martello and A. Smith (2014). "The Nature of Embryonic Stem Cells". *Annual Review of Cell and Developmental Biology* 30:647–675 (see p. 261).

156. Robert C. Martin, Dirk Riehle and Frank Buschmann, eds. (1998). *Pattern Languages of Program Design 3.* Addison-Wesley (see p. 13).

157. Wayne Materi and David S. Wishart (2007). "Computational Systems Biology in Cancer: Modeling Methods and Applications". *Gene Regulation and Systems Biology* 1:91–110 (see p. 63).

158. José L. Medina-Franco, Marc A. Giulianotti, Gregory S. Welmaker and Richard A. Houghten (2013). "Shifting from the single to the multitarget paradigm in drug discovery". *Drug Discovery Today* 18(9):495–501 (see p. 35).

159. Michelle C. Mendoza, E. Emrah Er and John Blenis (2011). "The Ras-ERK and PI3K-mTOR pathways: cross-talk and compensation". *Trends in Biochemical Sciences* 36(6):320–328 (see p. 35).

160. M. D. Mesarovic, S. N. Sreenath and J. D. Keene (2004). "Search for organising principles: understanding in systems biology". *Systems Biology* 1(1):19–27 (see p. 249).

161. Bertrand Meyer (2014). *Agile!: the good, the hype and the ugly*. Springer (see p. 241).

162. Olivier Michel (2004). "Webots: Professional Mobile Robot Simulation". *International Journal of Advanced Robotic Systems* 1(1):39–42 (see pp. 9, 30).

163. Alan Millard, Jon Timmis and Alan F. T. Winfield (2014). "Run-time Detection of Faults in Autonomous Mobile Robots Based on the Comparison of Simulated and Real Robot Behaviour". *IEEE/RSJ International Conference on Intelligent Robots and Systems (IROS 2014)*. IEEE, pp. 3720–3725 (see p. 37).

164. S. Muñoz Descalzo, P. Rué et al. (2013). "A competitive protein interaction network buffers Oct4-mediated differentiation to promote pluripotency in embryonic stem cells". *Molecular Systems Biology* 9:694 (see p. 261).

165. Egbert H. Van Nes and Marten Scheffer (2005). "A strategy to improve the contribution of complex simulation models to ecological theory". *Ecological Modelling* 185(2-4):153–164. DOI: 10.1016/j.ecolmodel.2004.12.001 (see p. 139).

166. J. Nichols and A. Smith (2009). "Naive and Primed Pluripotent States". *Cell Stem Cell* 4(6):487–492 (see p. 261).

167. Basarab Nicolescu (2010). "Methodology of Transdisciplinarity – Levels of Reality, Logic of the Included Middle and Complexity". *Transdisciplinary Journal of Engineering & Science* 1(1):19–38 (see p. 191).

168. G. Nicolis and I. Prigogine (1977). *Self-Organization in Nonequilibrium Systems*. John Wiley & Sons (see p. 261).

169. Michael Nielsen (2009). "Doing science in the open". *Physics World* 22(5):30–35 (see p. 40).

170. H. Niwa, Jun-ichi Miyazaki and A. G. Smith (2000). "Quantitative expression of Oct-3/4 defines differentiation, dedifferentiation or self-renewal of ES cells". *Nature Genetics* 24(4):372–6 (see p. 261).

171. Paul J. O'Dowd, Matthew Studley and Alan F. T. Winfield (2014). "The distributed co-evolution of an on-board simulator and controller for swarm robot behaviours". *Evolutionary Intelligence* 7(2):95–106 (see p. 37).

172. Howard T. Odum (1994). *Ecological and General Systems: an introduction to systems ecology*. revised edition. University Press of Colorado (see p. 126).

173. G. Parisi (1993). "Statistical physics and biology". *Physics World* 6:42–47 (see p. 262).

174. Amisha Patel, Nicola Harker, Lara Moreira-Santos, Manuela Ferreira, Kieran Alden, Jon Timmis, Katie Foster, Anna Garefalaki, Panayotis Pachnis, Paul Andrews, Hideki Enomoto, Jeffrey Milbrandt, Vassilis Pachnis, Mark C. Coles, Dimitris Kioussis and Henrique Veiga-Fernandes (2012). "Differential RET Signaling Pathways Drive Development of the Enteric Lymphoid and Nervous Systems". *Science Signaling* 5(235):ra55–ra55 (see p. 332).

175. Nikunjkumar Patel, Barbara Wiśniowska, Masoud Jamei and Sebastian Polak (2017). "Real Patient and its Virtual Twin: Application of Quantitative Systems Toxicology Modelling in the Cardiac Safety Assessment of Citalopram". *The AAPS Journal* 20(1):6 (see p. 36).

176. Fiona A. C. Polack (2010). "Arguing Validation of Simulations in Science". *Proceedings of the 2010 Workshop on Complex Systems Modelling and Simulation, Odense, Denmark*. Ed. by Susan Stepney, Peter H. Welch, Paul S. Andrews and Adam T. Sampson. Luniver Press, pp. 51–74 (see pp. 85, 190, 191, 327).

177. Fiona A. C. Polack (2012). "Choosing and adapting design notations in the principled development of complex systems simulations for research". *Proceedings of the Modelling of the Physical World Workshop*. ACM, p. 6 (see p. 73).

178. Fiona A. C. Polack, Paul S. Andrews, Teodor Ghetiu, Mark Read, Susan Stepney, Jon Timmis and Adam T. Sampson (2010). "Reflections on the Simulation of Complex Systems for Science". *ICECCS 2010*. IEEE Press, pp. 276–285 (see p. 327).

179. Fiona A. C. Polack, Paul S. Andrews and Adam T. Sampson (2009). "The engineering of concurrent simulations of complex systems". *CEC 2009*. IEEE Press, pp. 217–224 (see p. 327).

180. Fiona A. C. Polack, Alastair Droop, Philip Garnett, Teodor Ghetiu and Susan Stepney (2011). "Simulation validation: exploring the suitability of a simulation of cell division and differentiation in the prostate". *Proceedings of the 2011 Workshop on Complex Systems Modelling and Simulation, Paris, France*. Ed. by Susan Stepney, Peter Welch, Paul S. Andrews and Carl G. Ritson. Luniver Press, pp. 113–133 (see pp. 190, 191, 327, 329).

181. Fiona A. C. Polack, Tim Hoverd, Adam T. Sampson, Susan Stepney and Jon Timmis (2008). "Complex Systems Models: Engineering Simulations". *ALife XI, Winchester, UK, August 2008*. MIT Press, pp. 482–489 (see p. 7).

182. R. S. Pressman and B. R. Maxim (2014). *Software Engineering: A Practitioner's Approach*. 8th edition. McGraw-Hill (see p. 27).

183. Przemyslaw Prusinkiewicz, Scott Crawford, Richard S Smith, Karin Ljung, Tom Bennett, Veronica Ongaro and Ottoline Leyser (2009). "Control of bud activation by an auxin transport switch". *PNAS* 106(41):17431–17436 (see p. 212).

184. Steven F. Railsback (2001). "Concepts from complex adaptive systems as a framework for individual-based modelling". *Ecol. Modelling* 139(1):47–62 (see pp. 222, 224).

185. Mark Read (2012). "Statistical and Modelling Techniques to Build Confidence in the Investigation of Immunology through Agent-Based Modelling". PhD thesis. University of York (see pp. 181, 331).

186. Mark N. Read, Kieran Alden, Louis M. Rose and Jon Timmis (2016). "Automated multi-objective calibration of biological agent-based simulations". *Journal of The Royal Society Interface* 13(122):DOI: 10.1098/rsif.2016.0543 (see p. 164).

187. Mark Read, Paul S. Andrews, Jon Timmis and Vipin Kumar (2009). "A Domain Model of Experimental Autoimmune Encephalomyelitis". *Proceedings of the 2009 Workshop on Complex Systems Modelling and Simulation, York, UK.* Ed. by Susan Stepney, Peter H. Welch, Paul S. Andrews and Jon Timmis. Luniver Press, pp. 9–44 (see pp. 113, 188, 327).

188. Mark Read, Paul S. Andrews, Jon Timmis and Vipin Kumar (2014). "Modelling biological behaviours with the unified modelling language: an immunological case study and critique". *Journal of the Royal Society Interface* 11(99):(see pp. 113, 181).

189. Mark Read, Paul S. Andrews, Jon Timmis, Richard A. Williams, Richard B. Greaves, Huiming Sheng, Mark Coles and Vipin Kumar (2013). "Determining disease intervention strategies using spatially resolved simulations". *PloS one* 8(11):e80506 (see p. 330).

190. Howard Rheingold (1992). *Virtual Reality.* Simon and Schuster (see p. 8).

191. Angelika Riedl, Michaela Schlederer, Karoline Pudelko, Mira Stadler, Stefanie Walter, Daniela Unterleuthner, Christine Unger, Nina Kramer, Markus Hengstschläger, Lukas Kenner, Dagmar Pfeiffer, Georg Krupitza and Helmut Dolznig (2017). "Comparison of cancer cells in 2D vs 3D culture reveals differences in AKT–mTOR–S6K signaling and drug responses". *Journal of Cell Science* 130(1):203–218 (see p. 32).

192. Marylyn Ritchie, Emily Holzinger, Ruowang Li, Sarah Pendergrass and Dokyoon Kim (2015). "Methods of integrating data to uncover genotype-phenotype interactions". *Nature Reviews Genetics* 16:85–97 (see p. 38).

193. Frank E. Ritter, Michael J. Schoelles, Karen S. Quigley and Laura Cousino Klein (2011). "Determining the Number of Simulation Runs: Treating Simulations as Theories by Not Sampling Their Behavior". *Human-in-the-Loop Simulations.* Springer, pp. 97–116 (see p. 178).

194. Tânia Rodrigues, Banani Kundu, Joana Silva-Correia, S.C. Kundu, Joaquim M. Oliveira, Rui L. Reis and Vitor M. Correlo (2018). "Emerging tumor spheroids technologies for 3D in vitro cancer modeling". *Pharmacology and Therapeutics* 184:201–211 (see p. 33).

195. Eric Rohmer, Surya P. N. Singh and Marc Freese (2013). "V-REP: A versatile and scalable robot simulation framework". *Proc. IEEE/RSJ International Conference on Intelligent Robots and Systems, Tokyo, 2013.* IEEE, pp. 1321–1326 (see p. 30).

196. Andreas Rüping (2003). *Agile Documentation.* Wiley (see p. 217).

197. Derek Ruths, Melissa Muller, Jen-Te Tseng, Luay Nakhleh and Prahlad T. Ram (2008). "The Signaling Petri Net-Based Simulator: A Non-Parametric Strategy for

Characterizing the Dynamics of Cell-Specific Signaling Networks". *PLoS Comput Biol* 4(2):e1000005 (see p. 63).

198. Erol Şahin and Alan F. T. Winfield, eds. (2008). *Swarm Intelligence: special issue on Swarm Robotics.* Vol. 2(2–4) (see p. 240).

199. Andrea Saltelli, Karen Chan and E. Marian Scott, eds. (2000). *Sensitivity Analysis.* Wiley (see p. 177).

200. S. Sanders, A. C. Rolfe and EDFA-JET workprogramme (2003). "The use of virtual reality for preparation and implementation of JET remote handling operations". *Fusion Engineering and Design* 69(1–4):157–161 (see p. 11).

201. Robert G. Sargent (2005). "Verification and validation of simulation models". *37th Winter Simulation Conference.* ACM, pp. 130–143 (see pp. 7, 328).

202. Viktoria Spaiser, Peter Hedström, Shyam Ranganathan, Kim Jansson, Monica K. Nordvik and David J. T. Sumpter (2018). "Identifying Complex Dynamics in Social Systems: A New Methodological Approach Applied to Study School Segregation". *Sociological Methods and Research* 47(2):103–135 (see p. 41).

203. Ned Stafford (2010). "Science in the digital age". *Nature* 467:S19–S21 (see p. 40).

204. Susan Stepney (2013). "CoSMoS simulation experiment reproducibility and the ODD protocol". *Proceedings of the 2013 Workshop on Complex Systems Modelling and Simulation, Milan, Italy, July 2013.* Ed. by Susan Stepney and Paul S. Andrews. Luniver Press, pp. 93–107 (see pp. 223, 327).

205. Susan Stepney and Paul S. Andrews, eds. (2013). *Proceedings of the 2013 Workshop on Complex Systems Modelling and Simulation, Milan, Italy, July 2013.* Luniver Press (see p. 327).

206. Susan Stepney and Paul S. Andrews, eds. (2014). *Proceedings of the 2014 Workshop on Complex Systems Modelling and Simulation, New York, NY, USA, July 2014.* Luniver Press (see p. 327).

207. Susan Stepney and Paul S. Andrews (2015a). "CoSMoS special issue editorial". *Natural Computing* 14:1–6 (see p. 327).

208. Susan Stepney and Paul S. Andrews, eds. (2015b). *Proceedings of the 2015 Workshop on Complex Systems Modelling and Simulation, York, UK, July 2015.* Luniver Press (see p. 327).

209. Susan Stepney, Paul S. Andrews and Mark N. Read, eds. (2012). *Proceedings of the 2012 Workshop on Complex Systems Modelling and Simulation, Orleans, France.* Luniver Press (see p. 327).

210. Susan Stepney, Fiona Polack and Peter Welch, eds. (2008). *Proceedings of the 2008 Workshop on Complex Systems Modelling and Simulation, York, UK.* Luniver Press (see p. 327).

211. Susan Stepney, Peter H. Welch, Paul S. Andrews and Adam T. Sampson, eds. (2010). *Proceedings of the 2010 Workshop on Complex Systems Modelling and Simulation, Odense, Denmark.* Luniver Press (see p. 327).

212. Susan Stepney, Peter H. Welch, Paul S. Andrews and Jon Timmis, eds. (2009). *Proceedings of the 2009 Workshop on Complex Systems Modelling and Simulation, York, UK.* Luniver Press (see p. 327).

213. Susan Stepney, Peter Welch, Paul S. Andrews and Carl G. Ritson, eds. (2011). *Proceedings of the 2011 Workshop on Complex Systems Modelling and Simulation, Paris, France*. Luniver Press (see p. 327).

214. Chong Sun and René Bernards (2014). "Feedback and redundancy in receptor tyrosine kinase signaling: relevance to cancer therapies". *Trends in Biochemical Sciences* 39(10):465–474 (see p. 35).

215. J. Teles, C. Pina et al. (2013). "Transcriptional Regulation of Lineage Commitment – A Stochastic Model of Cell Fate Decisions". *PLOS Computational Biology* 9(8) :e1003197 (see p. 261).

216. John Timmer (2010). "Keeping computers from ending science's reproducibility". *Ars Technica* :arst.ch/d1p (see p. 7).

217. Jon Timmis, Kieran Alden, Paul Andrews, Ed Clark, Adam Nellis, Becky Naylor, Mark Coles and Paul Kaye (2017). "Building confidence in quantitative systems pharmacology models: An engineer's guide to exploring the rationale in model design and development". *CPT Pharmacometrics & Systems Pharmacology* 6(3):156–167 (see p. 191).

218. R. Turner (2007). "Toward Agile systems engineering processes". *Crosstalk* :11–15 (see p. 27).

219. Dieter Vanderelst and Alan Winfield (2017). "Rational imitation for robots: the cost difference model". *Adaptive Behavior* 25(2):60–71 (see p. 37).

220. Dieter Vanderelst and Alan Winfield (2018). "An architecture for ethical robots inspired by the simulation theory of cognition". *Cognitive Systems Research* 48:56–66 (see p. 37).

221. Marco Viceconti, Peter Hunter and Rod Hose (2015). "Big data, big knowledge: big data for personalized healthcare". *IEEE Journal of Biomedical and Health Informatics* 19(4):1209–1215 (see p. 36).

222. T. Vicsek (2002). "The bigger picture". *Nature* 418:131 (see p. 262).

223. John Vlissides, James O. Coplien and Norman L. Kerth, eds. (1996). *Pattern Languages of Program Design 2*. Addison-Wesley (see p. 13).

224. Christine E. Wania (2016). "Investigating an author's influence using citation analyses: Christopher Alexander (1964–2014)". *Proceedings of the Association for Information Science and Technology* 52(1):1–10 (see p. 14).

225. Christine E. Wania (2017). "Patterns and Pattern Sites in HCI: An Analysis". *SAIS 2017 Proceedings*. 9. URL: aisel.aisnet.org/sais2017/9 (see p. 13).

226. Colin Warwick (2009). "Everything you always wanted to know about SPICE* (*But were afraid to ask)". *EMC Journal* 82:27–29 (see p. 9).

227. D. J. Weatherall (2001). "Phenotype-genotype relationship in monogenic disease: lessons from the Thalassemias". *Nature Reviews Genetics* 2:245–255 (see p. 249).

228. Ken Webb and Tony White (2005). "UML as a cell and biochemistry modeling language". *BioSystems* 80:283–302 (see p. 113).

229. Herman A. van Wietmarschen, Heleen M. Wortelboer and Jan van der Greef (2016). "Grip on health: A complex systems approach to transform health care". *Journal of Evaluation in Clinical Practice* 24(1):269–277 (see p. 41).

230. Stephen P. Wilson, John A. McDermid, Clive H. Pygott and David J. Tombs (1996).
      "Assessing complex computer based systems using the goal structuring notation".
      *ICECCS 1996: Second IEEE International Conference on Engineering of Complex Com-
      puter Systems*. IEEE Press, pp. 498–505 (see p. 191).

231. John Wilson-Kanamori, Vincent Danos, Ty Thomson and Ricardo Honorato-Zim-
      mer (2015). "Kappa Rule-Based Modeling in Synthetic Biology". *Computational
      Methods in Synthetic Biology*. Ed. by Mario Andrea Marchisio. Humana Press,
      pp. 105–135 (see p. 113).

232. Alan F. T. Winfield, Christian Blum and Wenguo Liu (2014). "Towards an Eth-
      ical Robot: Internal Models, Consequences and Ethical Action Selection". *Advances
      in Autonomous Robotics Systems*. Ed. by Michael Mistry, Aleš Leonardis, Mark
      Witkowski and Chris Melhuish. Springer, pp. 85–96 (see p. 37).

233. Alan F. T. Winfield and Verena V. Hafner (2018). "Anticipation in Robotics". *Hand-
      book of Anticipation: Theoretical and Applied Aspects of the Use of Future in Decision
      Making*. Ed. by Roberto Poli. Springer, pp. 1–30 (see pp. 37, 38).

234. M. J. Wiser, N. Ribeck and R. E. Lenski (2013). "Long-term dynamics of adaptation
      in asexual populations". *Science* 342:1364–1367 (see p. 330).

235. H. Xu, C. Schaniel, I. R. Lemischka and A. Ma'ayan (2010). "Toward a complete in
      silico, multi-layered embryonic stem cell regulatory network". *Wiley Interdisciplin-
      ary Reviews: Systems Biology and Medicine* 2(6):708–733 (see pp. 260, 296).

236. P. Zandstra and G. Clarke (2014). "Computational Modeling and Stem Cell En-
      gineering". *Stem Cell Engineering*. Ed. by R. M. Nerem, J. Loring et al. Springer,
      pp. 65–97 (see p. 249).

Index

© Springer Nature Switzerland AG 2018
S. Stepney, F.A.C. Polack, *Engineering Simulations as Scientific Instruments: A Pattern Language*, https://doi.org/10.1007/978-3-030-01938-9

Printed in the United States
By Bookmasters